THE LIFE OF A NUMBER

Measurement, Meaning and the Media

B.T. Lawson

BRISTOL
UNIVERSITY
PRESS

First published in Great Britain in 2023 by

Bristol University Press
University of Bristol
1–9 Old Park Hill
Bristol
BS2 8BB
UK
t: +44 (0)117 374 6645
e: bup-info@bristol.ac.uk

Details of international sales and distribution partners are available at bristoluniversitypress.co.uk

British Library Cataloguing in Publication Data
A catalogue record for this book is available from the British Library

ISBN 978-1-5292-2533-4 hardback
ISBN 978-1-5292-2534-1 ePub
ISBN 978-1-5292-2535-8 ePdf

Cover design: Qube Design Associates, Bristol
Front cover image: shutterstock_2202144201
Bristol University Press use environmentally responsible print partners.
Printed and bound in Great Britain by CPI Group (UK) Ltd, Croydon, CR0 4YY

FSC www.fsc.org MIX Paper | Supporting responsible forestry FSC® C013604

To friends, family and colleagues that made this book more than the writings of one person.

To Katie, who helped me to stay sane.

Contents

1

Introduction

The coronavirus pandemic has resulted in a deluge of data: deaths from coronavirus, viewing figures for conspiracy theory videos, unequal vaccine coverage, changes in Gross Domestic Product (GDP), the list goes on. While much of this data has been useful in understanding and dealing with COVID-19, there has been a noticeable excess. This has flowed into the so-called 'post-pandemic' world, where numbers burst from the seams of public discourse. Experts constantly update us with newfangled metrics, politicians point to the latest iteration of international league tables and journalists report on a dizzying volume of data. We need to step away from this daily churn of the quantitative and ask: what does this data actually *mean*?

Some approach this effort as a technical exercise. During the pandemic, Tim Harford's *More or Less* show on BBC Radio 4 provided an excellent weekly dive into salient numbers. The team applied statistical rigour to certain factoids, helping them to detect the pitfalls of small sample sizes, the role of nefarious categorization for political ends and how experts would communicate misleading conclusions from the data (see *More or Less* (2021) for an example). This was undoubtedly important work. It allowed the public to navigate the sea of data that was flowing their way.

But it often erred on the side of 'if only they conducted the right statistical test, these numbers would not be a problem'. This meant that they missed a certain something about how numbers gained meaning in society. Quantitative facts are not just the result of mathematical and statistical processes. They are characterized by complexity and paradoxes: at once scientific and ideological, empowering and discriminatory, precise and uncertain, objective and subjective, emotive and informative, truthful and deceptive, fixed and malleable.

To untangle this complexity, I argue that we must begin by accepting two premises. First, we need to pay attention to the way mathematics and statistics *combine* with politics, culture, technology and economics. The *technical* process of producing a number – collecting data, cleaning it and analysing it – cannot be divorced from the *context* within which this occurs.

Politicians will demand that specific numbers are created to demonstrate the success of their policies, experts will work with private companies to research topics of corporate interest and data journalists will explore vast banks of public data to hold power to account.

Second, we cannot understand this relationship without media and communication. Social media allows politicians to frame their policies as successful through public communication. Television news channels facilitate experts' discussion of the latest findings from their research. Print newspapers provide a space for journalists to argue for the importance of particular datasets for democracy. Accepting both premises means seeing how the quantitative involves the meeting of the technical and context within media ecosystems.[1]

This sits at the heart of the Life of a Number approach, where seven numbers from the pandemic are analysed through a five-stage process: need, production, communication, public meaning and background. The story of each number forms the six empirical chapters of this book, each providing a self-contained story of how different parts of the quantitative became meaningful. When these stories are brought together, however, they point to a broader – more important – academic contribution.

Data bound

Data bound concerns how the quantitative becomes a meaningful way to experience, think about, discuss, react to, engage with, and change certain phenomena.[2] To outline the nature of data bound, we can see the term as both a verb and a noun: it is both something that is happening and something that exists.

Let's begin with the verb. We can see how data *bounds* a phenomenon by *representing* it. The pandemic, for example, was bound by the different numbers that were used to talk about it – this includes cases, hospitalizations, deaths, GDP, hospital waiting lists, vaccine efficacy, mobility data, and so on. To arrive at these numbers involves the quantification of reality: taking the messy world we live in, determining categories and classifications, using technology to collect and analyse data, and communicating specific league tables, indicators and statistics. This is not a static process; it is a continuous one where the form of quantification (the way things are categorized, the analysis chosen, and the variables examined) is constantly reaffirmed through the incessant production of data.

This process results in the noun: the *bounds* of the phenomenon. Certain ideas and actions are 'in bounds' (or logical and acceptable) and others are 'out of bounds' (or illogical and abhorrent). When we consider COVID-19, it was considered illogical and abhorrent that the government would do nothing in the face of rising cases or the threat of mass unemployment.

While the exact nature of this action was contested, the need to act was taken for granted.

Herein we see the power of data bounds to maintain specific versions of phenomena and marginalize others. In part, the ability of data to form these bounds is due to the trust we place in numbers to represent the world around us objectively and accurately. This belief system – called quantitative realism – gives these boundaries an impenetrability that the qualitative cannot.

But the creation, establishment and continued existence of data bounds stretch beyond trust in the quantitative. They are the product of the complex interactions between the state, economics, culture, history and media. Some of this is structural – such as the way the state facilitates capitalism – but a significant part is also strategic: data bounds are needed to exercise power, change society, suppress opposition and liberate the marginalized.

One way to untangle this web is to focus our attention on specific numbers that exist inside and outside data bounds during the pandemic. In mapping out the need for numbers, their creation, their communication and how they circulated in society allows us to take a deep-dive into COVID-19 data bounds through a single case study. The nuances of these stories can provide insights to data bounds as a whole and, most importantly, ways that they can be explored critically as scholars. This is how the six empirical chapters are framed in this book – with the six stories outlining six characteristics of data bounds:

1. Data bounds are reinforced by policy.
2. Quantitative realism underpins data bounds.
3. Quantitative realism is mathematical and abstract.
4. Desire for data bounds underpins quantitative realism.
5. Data bounds are emotive.
6. Data boundaries are drawn within historical norms.

These six perspectives on data bounds are brought together in the concluding chapter to form a four-piece toolkit for scholars looking to use the concept in their work:

1. Pay attention to media and communication.
2. Interrogate and appreciate quantitative realism.
3. Examine how data bounds can maintain or challenge power.
4. Determine why some data bounds dominate over others.

This toolkit is great for understanding the pandemic. But its utility stretches beyond COVID-19. It can be deployed on other highly quantified phenomena, such as health and fitness, inflation and cost of living, crime and justice, public opinion, risk (and many more). Such a task is even more important when we consider the ever-expanding net of data, characterized by the proliferation of smart devices in the home, the colossal funding for the Metaverse and the increasing capacity of supercomputers. As we *bound* towards even more data, scholars need to embrace data bounds.

Chapter 2: Data bounds are reinforced by policy

Chapter 2 focuses on two data bounds from the pandemic: *Protect Both* and *Trade-Off*. These provide two contradictory ways of understanding COVID-19. *Protect Both* argued that both the economy and public health could be protected if cases of coronavirus were kept very low (or at zero). *Trade-Off*, on the other hand, positioned public health in opposition to the economy – where national lockdowns and 'opening up' society meant countries shifted from high case numbers and good economic performance to low case numbers and poor economic performance. Despite being contradictory, both realities made sense when set in their own data terms.

Protect Both was comprised of international comparisons of macroeconomic and public health data. This pointed to a clear conclusion. Countries taking an elimination or containment strategy protected both lives and livelihoods, whereas nations opting for mitigation strategies saw larger losses of life *and* economic hardship.

Trade-Off used the same set of data as *Protect Both* but focused on short-term changes in public health and economic data. Seven-day rolling averages that showed rising cases, hospitalizations and deaths would underpin the need to enter a lockdown. Subsequent reductions in cases, hospitalizations and deaths would emphasize the need to emerge from a lockdown. This pattern repeated itself for the first, second and third lockdowns – resulting in stay-at-home orders for a total of just over half a year.

This highlights how data cannot 'speak for itself'. Rather, it was government policy that dictated the type of data bounds. The policy adopted in England meant the sentiment of 'obviously the government need to lockdown, look at the rising number of deaths' displaced the idea that 'the government's adherence to mitigation has cost needless lives'. In this way, *Trade-Off* involved a reinforced cycle between the data and policy. Each time the government introduced a lockdown, it reinforced the idea that the cycle of lockdowns–opening up–lockdowns needed to occur.

Chapter 3: Quantitative realism underpins data bounds

Chapter 3 examines the history of 'close contact' – defined as being within *two metres* of an infected person for *15 minutes or longer*. The chapter argues that these two precise figures provide an illusion of certainty to the incredibly complex and context-dependent phenomena of transmission.

By tracing the history of both numbers, the chapter puts forward a three-part explanation of how numbers render the unknown 'knowable': (1) the *language* of mathematics deals in universal absolutes; (2) this language is materialized through *measurement* devices; and (3) *documents* take these numerical measurements, combine them with other forms of representation, and create immutable and mobile scientific documents.

This process of language, measurement and documentation underpins the paradigm of *quantitative realism*: that the world *can be* quantified and, through quantification, *reveals* itself.[3] Such a notion is entrenched in modern society – most people fully accept that almost everything can be represented through numbers and that these numbers hold a deep truth about a phenomenon.[4] The dominance of *quantitative realism* in contemporary society allows numbers to construct data bounds.

The certainty of the quantitative means that data bounds have a rigidity and impenetrability that other realities do not. Saying to someone that you should not go near another person for a long period of time links into our 'common sense' understandings of how diseases are spread. But providing two numbers – two metres and 15 minutes – forms the girders that rigidly fix transmission. So, we need to see how *quantitative realism* underpins the power of data bounds.

Chapter 4: Quantitative realism is mathematical and abstract

Chapter 4 shifts from an exclusively science-meets-mathematics understanding of quantitative realism to incorporate abstraction. It focuses on government officials' use of 'one billion items of PPE [Personal Protective Equipment]' to distract and deflect from the consistent complaints by frontline healthcare workers in the early stages of the pandemic that their equipment was inadequate (Price and Harbisher, 2022). The number did not just serve to hide immediate failings, but longer-term ones too: lack of adequate stockpiles, inadequate production and procurement networks, prioritizing the National Health Service (NHS) over social care, and changes in what classifies as suitable PPE.

A particularly powerful element of this rhetoric was the 'hugeness' of one billion itself. It conjured a notion of largeness that matched with the idea that they were 'doing everything they could' in the face of an 'unprecedented crisis' to secure and provide PPE to workers. But looking more closely at the figure, we can see that this 'hugeness' was an artifice. Imaginative accounting meant that 'one billion' was comprised of 500 million individual gloves and thousands of bottles of bleach. The number defied both logic and official definitions of PPE.

Herein lies two important lessons for quantitative realism. Large numbers can occupy a position of mathematical certainty *and* an abstract, non-mathematical notion of 'hugeness'. This is especially true when numbers reach past a certain scale, stretching into the millions, billions and trillions. The power of quantitative realism is showcased in the ability of numbers to remain innocent until proven guilty. The figure of 'one billion' was treated as a mathematical and abstracted 'fact', despite the fact it emerged from a deliberate manipulation of the dataset describing PPE.

Chapter 5: Desire for data bounds underpins quantitative realism

Chapters 3 and 4 have emphasized the way quantitative realism has become the dominant paradigm in society by focusing on numbers. But the need for data bounds can also underpin a belief in quantitative realism. In other words, the power of data bounds makes quantitative realism a necessary achievement.

To outline how this occurs, Chapter 5 focuses on *the infodemic* – a second-tier, more narrowly focused data bound that emphasizes how the spread of misinformation (predominantly online) is leading to mass non-conformist behaviour that poses a considerable risk to public health. Underpinning the infodemic was a host of data: the number of posts flagged as 'misinformation' on social media platforms, the percentage of people who believed in conspiracy theories, the level of compliance in different countries, and so on.

But, as the chapter goes on to highlight, the most robust empirical data showed that the presence of mass non-conformist behaviour was actually incredibly low during the early stages of the pandemic. Movement, survey and national transport data all pointed to high levels of adherence to lockdown rules in England (Anderson, 2020a; Jarvis et al, 2020). Nevertheless, the infodemic continued to exist as the predominant way of conceptualizing misinformation – emphasizing a considerable threat to public health.

How can this be explained? This chapter argues that the news media needed the infodemic to position themselves as the antidote: we are trusted gatekeepers of credible information that informs our audience, not an unregulated space of misinformation weaponized to dupe the public (Carlson, 2018). To maintain this data bound, the same journalists needed to believe in the quantitative realism underpinning it. To outline how this interplay worked, the chapter looks at two numerical cases in US and UK news media discourses: '8 million people have watched *The Plandemic*' and '7% of people believe there is no hard evidence for COVID-19'.

When we engage with this type of quantification, we need to be careful. It is easier to count the number of cases of COVID-19 or individual items of PPE than measure the beliefs, desires and intentions of human beings. This is highlighted in a breakdown of both statistics: the 'eight million' statistic involves an oversimplification of 'the view' as representing people's belief in the *The Plandemic* conspiracy video that was watched, whereas the 7 per cent figure gives too much certainty to the answer to a single question during a survey by Kings College London and Ipsos Mori in understanding human beliefs and intentions (Moon, 1999; Cheney-Lippold, 2017; Lupton, 2019).

Despite the shortcomings of this data, journalists needed to believe and assert its quantitative realism to allow the infodemic to function. In doing so, they did construct their own identity as trusted gatekeepers of information but also put forward a version of reality that was misleading. It shifted the blame for rising cases, hospitalizations and deaths away from political decisions towards the misinformed public.

Chapter 6: Data bounds are emotive

Chapter 6 emphasizes how people's relationship to numbers extends beyond the conventional ideas of rationality, knowledge and calculation. Instead, data bounds involve emotional experiences, characterized by strong feelings attached to data.

Whereas other chapters have centred on numbers that are represented through text or verbal communication, Chapter 6 focuses on the importance of data visualizations. The case study centred on the data visualizations used in a Sky News YouTube clip from November 2020 titled *How did the UK get to 50,000 deaths?* In this clip the presenter – Roland Manthorpe – uses the image of the number of deaths per day over time. This peak and trough graph became iconic during the pandemic.

But a graph representing death is not death itself. There always involves a process of abstraction. The chapter outlines how data visualizations involve *selection*, *simplification* and *mathematization* to establish quantitative realism during this process of abstraction. Given that these three stages occur across data visualizations, what made *this* data visualization special?

The chapter emphasizes that the data visualization was set within the *Trade-Off* data bound described in Chapter 2, with the peaks and troughs of daily deaths representing the shift from lockdown to opening up to lockdown to opening up. But the visualization seemed to do specific work within this reality in capturing a collective experience of these policy decisions: mass changes in behaviour, the trauma of having family and friends contract, be hospitalized or die from the virus and a cultural memory of how the pandemic progressed through 2020 to 2021.

This meant the graph had an emotional investment. This is captured in the performance of Roland Manthorpe when he presents the visualization in his YouTube clip. He seems to overstep an invisible line in the genre of presenting data visualizations about death. His excitable, expressive and dramatic performance sits at odds with a data visualization that seems to capture the collective experience of the pandemic. In doing so, the chapter emphasizes the way data bounds are experienced and become felt.

Chapter 7: Data boundaries are drawn within historical norms

The previous five chapters are about numbers that *did* circulate: the definition of close contact (two metres and 15 minutes), macroeconomic indicators and public health metrics, one billion items of PPE, the peak and trough data visualization and statistics concerning people's belief in misinformation. But Chapter 7 is about a number that *did not* circulate. It focuses on a simple projection made by Professor Christina Pagel, a Professor at University College London, at a weekly YouTube Independent SAGE briefing.

She claimed that "in mid-July [2021] we will have a seven-day average of 90,000 [cases per day]". This underpinned her argument that the government should not end restrictions on 19 July 2021 as planned (so-called 'freedom day'). This figure, however, received very little coverage outside the Independent SAGE weekly briefing.

The chapter examines why the projection failed to take off. Some argued that it was ignored because it was a poor projection. This proved to be the case, with the figure considerably overestimating the scale of cases. But this ignores other poor projections that gained a lot of attention. In search of an answer, the chapter conceptualizes the projection as a form of risk communication. The risk of 90,000 cases per day was seen as an acceptable trade-off for ending most of the restrictions in England. But how can the government accept such a risk?

First, we need to consider how this discussion of risk was set within the pandemic-specific data bound of *Trade-Off*. The 19 July 2021 – 'freedom day' – was largely about achieving near-complete economic freedom for the population as both consumers (the freedom to spend money in shops, restaurants and bars) and as workers (the freedom to go back to work and generate money). It was hoped that this economic freedom would boost the macroeconomic indicators. The trade-off involved a public health risk of rising cases, hospitalizations and deaths.

The chapter argues that this risk was acceptable because the rising cases disproportionately affected the most deprived in society. It was this group that would experience more cases, more hospitalizations, more deaths, more missed school and work days and more pressure at their local hospital. Quite simply, the risk to the most deprived was acceptable *because* the risk to everyone else was acceptably low. To understand why this was the case, we need to look beyond *Trade-Off* towards the historical normalization of health inequalities.

This is the culmination of a long process, one that this chapter traces from the 19th-century Victorian period through to austerity in the 2010s. Within this norm, it was seen as absurd to *not* open up society when most of the risk of the virus would only be felt by the most deprived people. Therefore, the '90,000 cases' projection failed to take off because of the implicit logic within *Trade-Off* that unequal health risks were acceptable. A logic that is rooted in a centuries-long normalization of health inequalities.

Chapter 8: Critically engaging with data bounds

The six main chapters point to a different perspective of data bounds. The final chapter looks to step beyond the characteristics, towards a guide for how to expertly explore them. It argues that scholars must do four things:

1. *Pay attention to media and communication*: Through an analysis of media systems, scholars can understand how data is given meaning in society and, in the process, given the power to establish what is within data bounds and what is outside. To do so, attention must be paid towards the different components of the media ecosystem: the forms of communication, the spectrum of meaning, the types of media and the communicators. It is through placing media and communication front-and-centre that sets data bounds apart from the range of other data-prefixes in academia (for example, data infrastructure, data engines, data assemblages, and so on).
2. *Interrogate and appreciate quantitative realism*: Engaging in this dual task is of vital importance. Quantitative realism cannot be taken for granted – scholars must critically engage with the affordances and limitations of quantification from both a technical and philosophical position. But an appreciation must be given to the power of quantitative realism as an overarching way of thinking – and how this power is wielded by experts, politicians and journalists.
3. *Examine how data bounds maintain or challenge power*: There are many possibilities within data bounds, three of which are identified for the reader. They can serve a public health imperative that helps central authority to manage crises, they can function to expose political ineptitude and highlight viable alternatives and they can become a political requirement to maintain the status quo. How data bounds relate to power – and the purpose of this power – sits at the heart of their significance.
4. *Determine why some data bounds dominate and others do not*: Four explanations provide a starting point for mapping out the factors that allow some data bounds to out-compete others. The first concerns *political consensus*, the second *historical context*, the third *experience* and the fourth the *runaway train of meaning* that some data acquires.

More cases, less theory

As the overview above has emphasized, each chapter provides a theoretical contribution to data bounds. They help the reader gain multiple complementary perspectives on the concept. But the book did not start from the point of theory. Each chapter began as a case study of one or two particularly interesting numbers from COVID-19, analysed the life of each figure and constructed a narrative about them. Each stage was informed by theory but driven by the case study. It was only afterwards that the broader theoretical contribution emerged.

Some would see this methodology as being too specific, unable to be replicated, ungeneralizable and too open for my biases. These people would be more at home with standard research designs that involve a larger sample size and a combination of interpretive and statistical methods. This book does not disregard the importance of these types of analyses, rather it follows the call of Berman and Hirschman (2018) to trace a small number of quantitative cases in detail.[5] And such an approach is not unprecedented. A case study approach can be observed across literature from different disciplines.

When dealing with mathematics itself, work from the history and philosophy of numbers will often focus on how certain fundamentals of mathematics emerged. Clawson (1994) provides a whole chapter dedicated to the concept of π, while Higgins (2008) gives substantial space to the notion of infinity. Beyond mathematical concepts, other work has focused on how certain quantitative concepts about the social world were established in the 19th century. Most notably, the academic work on Adolphe Quetelet's *L'homme Moyen* (1844–48) who set out the case for seeing the social world as governed by *averages* and *rates* (Curtis, 2002).

Outside of a history and philosophy of mathematics and statistics, it is more common for quantification to be set in relation to politics, the social, discourse, institutions, practices and actors. This often emerges in the way of anecdotal examples by authors to emphasize the particular power that numbers afford in governance. In his 2001 paper, Peters (2001) points to the counting of slaves in the 18th-century United States. To determine the number of seats that a particular state would have in the US House of Representatives for the next ten years, the population of each state needed to be determined. Given that there was a very large slave population in the southern states, there was a push by these states to count these slaves in their population figures. But a slave could not be counted as a non-slave as this would contradict their enslaved status. Therefore, the 1787 US constitutional convention agreed on the 'Three-Fifths Compromise': for every five slaves, three should be counted. In other words, slaves were treated as three-fifths of a person (Peters, 2001).[6]

But these cases do not just serve as anecdotal examples. Some scholars have made specific metrics the focus of their work. Probably the most comprehensive account comes from Espeland and Sauder (2016) when they examine the way law school rankings have changed legal education in *Engines of Anxiety*. Turning to economics, the work of Fioramonti (2013, 2014) on GDP has helped to historicize this macroeconomic indicator, highlight its technical deficiency and question its relevance in the 21st century. But the focus is not only on contemporary forms of quantification. Deringer (2018) paid close attention to the £398,085 10s. (or the 'Equivalent') – the amount of money paid by the English government to the Scottish to secure them joining the union.[7]

Precedent is not everything. What exactly does a case study approach provide that others cannot? Michael Billig's book *More Examples, Less Theory* (from which the subheading for this section was derived) (Billig, 2019) is instructive to much of the *why* for case studies.

Principally, they are led by the example rather than the theory. This pushes against work that starts with a particular theory about numbers, such as 'quantitative governance', and approaches examples through this theoretical framework. The resulting analysis can often feel like the author has found examples to support their particular theoretical position, even if this is not the methodological tack they have taken. Most often this emerges in the use of vague, overly broad language that – at its worse – loses the reader in phrases that describe so much, it ends up saying nothing at all.

In its place, the case study approach lets the individual example do the leading. After spending a considerable amount of time with the case study – and subjecting it to analysis – it will often point to relevant theory. This theory is then used to think through the example. In the words of Michael Billig (2019, p 236), 'theory is ousted from its position of command, so examples fill the vacated space'.

Beyond this, case studies have a particular power to operate as metonyms – where the number itself can stand in for a broader associated phenomenon. As Michael Billig (2019, p 228) argues, 'vivid use of examples enables a particular case to stand for a much greater whole'. This stands somewhat at odds with the logic of representation we can see in quantitative methods where, for example, 2,500 people are selected to be representative of the UK population in a survey. The case study approach does not say 'this is statistically representative', it says 'this gets at a broader truth, trend or pattern'.[8] As Michael Billig (2019) argues through the work of Jahoda's *Marienthal* (1933), it was a specific story, rather than a statistic, that provided the best account of 'the communal spirit of Marienthal and the simultaneous threat that poverty makes to that spirit'. Jahoda et al (2002, p 22) writes that 'when a cat or a dog disappears, the owner no longer bothers to report the loss; he knows that someone must have eaten the animal, and does not want to find out who'.

In part, the ability to operate as a metonym is also a result of being driven by the case and not the theory. It allows for a narrative that anchors itself in something familiar and explained in accessible language, rather than the complex jargon that academia is well known for. It is not to say this jargon is rejected entirely. Rather, it is introduced when needed and often after setting the scene of the case study. This allows non-specialists to engage in the work but also affords a level of intelligibility to academic readers who do not come from the disciplines that influence this book.

The life of a number

Now the merits of the case study approach have been fleshed out, there is a need to outline exactly *how* it was done. Methodology sections in academic books have become increasingly unfashionable. Outlining the *how* is seen as something specific to the formal structure of journal articles that require an introduction, literature review, methodology, findings and concluding discussion. As this book is advocating for a method as much as an argument that results from the method, it dedicates some much-needed space to explaining how the case studies were selected.

This project is rooted in the PhD that I completed at the University of Leeds from 2017 to 2020, but its empirics can be traced back to March 2020. As the pandemic swept across Europe, and the UK introduced an unprecedented national lockdown, I began to record and explore particularly interesting numerical case studies. I had found this case study approach useful for my thesis (which was being written up at the time) and wanted to apply it to the pandemic as well. This process of selection was anything but formal – I noted down numbers I saw on the news, figures I came across in scientific documents, data mentioned by friends and family over the phone and, in my brief trips out of the house, quantitative information spread across brick walls, lampposts and the outside of cafes, shops and restaurants. My only criterion was whether the number struck me – either in its extraordinary nature or its banality.[9] Despite setting this barrier relatively high, I was soon inundated with different figures – amassing around 100 different case studies from January 2020 to July 2021.

This process of selecting numbers will not be to the liking of certain social sciences, but is a deliberate attempt to incorporate my experience of the pandemic into my writing about COVID-19. Everyone who studies COVID-19 has been affected by how their lives changed during the pandemic. And, I strongly believe, this experience of the pandemic then structures how they frame, conduct and narrate their research. The pandemic is not an abstract, historical or spatially removed phenomenon to study, it is lived one. Living in Leeds, Keighley and then Sheffield (all towns and cities in the UK), I experienced either a national or local lockdown from March 2020 to April 2021 – having no friends or family legally allowed to enter either of my two homes. My friend, sister, brother-in-law and next-door neighbour had positive tests in the run up to the October 2020 lockdown.

My mum and step-dad had to cancel funerals and weddings at their church. I attended a wake in October 2021 for a family friend who died in March 2020. And, as we were being told that the pandemic was coming to an end in March 2022, I tested positive for coronavirus. All of this is to say that the subjectivity of my work cannot be erased. But it can be organized.

With these 100 plus numerical cases about the UK, the issue of selecting specific examples emerged. This necessitated some sort of selection process to whittle down the case studies to a manageable number. It is here where my approach became more formal. I developed a matrix of variables to classify and order my case studies. This drew on my experience of dealing with numbers from my PhD but also my specific experience of the pandemic itself. This resulted in the following categories:

- **Month:** This referred specifically to what month the number emerged in public discourse.
- **Source:** This focused on where the number came from – a politician, expert or journalist.
- **Type of number:** This used a broad classification system developed during my PhD that split quantitative information into measurements, statistics, pledges or targets, index, ranking or league table or non-metrological (where the number is not derived from measurement, for example, '7up').
- **Temporal dimension:** This recorded whether the number referred to something in the past, present (or recent past) or future.
- **Phenomena being quantified:** The thing actually being quantified was split into broad categories of population, disease, economics, response, public behaviour and opinion, and time and space.
- **Representation:** This was principally concerned with how the number was presented: linguistic (written or spoken), visual (for example, data visualizations) or other, less common, forms of representation (including sound, smell and taste).
- **Causality:** This was focused on whether the number referred to one thing causing something else, for example, vaccination rate causing a reduction in transmission of the virus.
- **Information or misinformation:** To be classified as misinformation, the number would need to either be (a) a figure that was produced or presented in a misleading manner that fit the idea of misinformation or disinformation and/or (b) an accurate, reliable and valid number that described misinformation (either circulating online, people's beliefs, and so on).

From this matrix, I then selected seven case studies that could tick as many of these boxes as possible. Once I had my cases, I needed a somewhat systematic way to dive deep into each case study. To do so, I developed the 'Life of a Number' methodological approach – comprised of five parts:[10]

- **Need:** If there is a 'beginning' to a number, then this is where we can find it. It is only when someone or something needs a number to exist that such a figure can emerge. This need can come from a politician looking to score political points, an expert looking into a phenomenon, or even a piece of artificial intelligence.
- **Production:** It is no good just demanding for data to exist, it needs to be produced through a series of technical processes. This includes a range of activities: defining and categorizing the phenomena in question, collecting data, analysing it and interpreting it.
- **Communication:** Once the data is produced, there is almost always a direct act of communication about the data in question. This can be the politician emphasizing the level of poverty in society through a television interview, an expert putting out a press release via their university's website or a journalist producing a data journalism piece.[11]
- **Public meaning:** From this specific act of communication, the number enters the much broader, messy and complex world of public meaning. The figure may come from a university website, be tweeted by an academic, retweeted by an influencer, gain a considerable number of interactions, be picked up by a mainstream news channel and result in the same expert who produced the number doing an interview. This is just one variation of a countless set that could occur.
- **Background:** At some point, the number falls from public meaning. This could be when a newer number replaces it (for example, the way daily updates on coronavirus cases replace the previous day's count), when that number is found to be fabricated or when the figure is simply deemed uninteresting.[12]

While the narrative given above provides a start-to-finish or linear explanation, the life of a number is anything but. We can consider a hypothetical situation. A set of data is **produced** by a research group at a university, and it gains a significant amount of attention in **public meaning**. In response, the vice-chancellor of said university explains the **need** for similar research to be conducted so the university maintains its elevated profile. The new data is **communicated** to a wider audience and nobody is that interested, so the number sinks into the **background** almost immediately after it emerged. But then, four years later, a specific event happens that draws attention to this research again and, without the **communication** of the experts, it is thrust back into **public meaning** again. Beer refers to these sorts of processes as feedback loops – characterized by repetition, constant iteration and change (Beer, 2016).

This book's application of the Life of a Number is led by stages three and four. That is, it starts from the point at which a number is communicated and emerges in public meaning. From this point, the story travels both forward and backwards. It traces the measurement of the figure, as put outlined in the need and production parts of the approach. But it also goes forward, by seeing when (and, in some cases, if) the number receded into the background.

In part, you can locate my emphasis on communication and meaning in my training as a media and communications scholar. Much of this work begins by seeing when, how and why information, ideas, practices, histories and mythologies circulate across television, radio, print and digital media by journalists, politicians, experts, the public, celebrities and other powerful individuals. And it is here that my fascination with numbers lies: in the way they are communicated and gain meaning across society in odd, banal, predictable and unforeseen ways. The realization of this comes with the Life of a Number approach.

The pandemic in the UK

Despite this systematic approach, the focus on the UK needs some defending. While it highlights my own entrenched nationalism, it should not undermine the book as a whole. The UK was a nation of extremes during the pandemic. But this does not mean that the book is just about the UK; many of the case studies span across national boundaries or speak to international issues. In this way, the chapter uses the UK as an anchoring point to explore intersections of numbers, communication and COVID-19 that resonate internationally. To appreciate the importance of looking at the UK, we need to map out a brief history of the pandemic in the UK.

To understand this history, we first need to appreciate that the UK is made up of four nations: Wales, Scotland, Northern Ireland and England. While fiscal policy – money given to each nation by The Treasury – is centrally managed by the UK government, public health decisions are devolved. This currently means that the Conservative Party decides health policies in England, the Welsh Labour Party in Wales, the Scottish National Party in Scotland and the Northern Ireland Executive in Northern Ireland. Such a set-up resulted in public health measures differing across the four nations. A prime example of this in the pandemic was Scotland's attempt at zero COVID during the summer of 2020, while England looked to manage relatively high case numbers in society. In certain key moments, however, the four nations implemented COVID-19 policy at the same time – such as the first UK-wide national lockdown in late March 2020. When this book refers to public health policies, it is focusing on the Conservative Party's decisions about England.

Even though the virus was probably spreading before the first official case of SARS-CoV-2 was declared at the end of January 2020, other than calling for the public to wash their hands for 20 seconds with hot soapy water, very little changed in terms of official English government policy between the detection of this case and mid-March 2020.

This was despite widespread lockdowns in East Asia and, closer to the UK, the growing number of hospitalizations and deaths in mainland Europe in early March 2020. On 12 March, the Conservatives announced that the UK was moving into the 'delay phase', which included seven-day isolation for those with a positive test (alongside existing handwashing advice). On 16 March, the government announced they were transitioning to 'suppression'.

Over the next week, the Conservatives brought in a phased set of measures, resulting in the first UK-wide lockdown being implemented on 23 March, including seven-day isolation for anyone with a positive test, 14-day family isolation, closure of schools and creches for most children, stay-at-home orders for everyone who was not an essential worker and the closure of most indoor public spaces (for example, cafes, bars and gyms). Because the Conservative government were comparatively late to lockdown compared to their European neighbours, England and Scotland witnessed one of the highest excess deaths rates across the continent in the first half of 2020 (ONS, 2020b).

By June 2020, the number of coronavirus cases had been suppressed enough for England to emerge from these lockdown measures. But certain areas were kept in local lockdowns, such as Leicester, with other areas emerging from a national lockdown only to be placed back into a local lockdown, such as Bradford. Despite this, the summer was characterized by an emphasis on 'freedoms' to see people, to socialize, to go to work and to go on holiday – perhaps best epitomized by the Conservative Party's 'eat out to help out' scheme that offered two-for-one meals at certain restaurants, take-aways and pubs during August 2020.

But the summer of 'freedoms' meant that autumn 2020 witnessed rising cases. This forced the government to implement a second national lockdown at the start of November. Less restrictive than the first, these restrictions lasted for around one month. The emergence of the Alpha variant in early winter 2020 meant the four nations were placed into lockdown measures before Christmas – effectively 'cancelling Christmas'. A country-wide lockdown was announced in early January 2021.

By late 2020, the effect of the Conservatives' policy decisions were starting to emerge. Lurching from lockdowns to freedom to lockdowns (called a 'mitigation' approach) had resulted in a high number of deaths and poor overall economic performance compared to other countries that took an elimination (zero cases of coronavirus) or a containment strategy (keeping cases manageably low) (see Chapter 4 for more details). But at this time, the UK also set itself apart in relation to vaccines. It was the first country in the world to approve a vaccine for emergency use in early December (Baraniuk, 2021). This meant that the UK was the first of the G20 nations to vaccinate over 50 per cent of adults with a first dose by late April 2021 (Cherry and Capel, 2021).

This successful rollout meant that the risk from coronavirus had changed: it was harder for the virus to spread and, if someone contracted the virus, they would be less likely to be hospitalized and die. Therefore, as the Delta variant emerged in spring 2021, England entered into a unique race between the rollout of the vaccination programme and exponentially rising cases of the new strain of the virus (Sample, 2021). Such was the optimism within government that they announced 'Freedom Day' on 19 July 2021 – where most of the existing restrictions were repealed.

The summer to autumn period of 2021 was characterized by a steadily growing number of daily cases but relatively little pressure on the hospital system and much lower death rates (compared to previous waves). As the country headed towards the second Christmas of the pandemic, the threat of the Omicron variant led the government to introduce their 'Plan B': introduction of masks in public spaces, restrictions on foreign travel, advice to work from home and to reduce social contacts. Despite calls for more stringent measures, the English government stood firm in their approach throughout January 2022. By 26 January, the government had repealed the measures involved in Plan B.

2

Data Bounds Are Reinforced by Policy

As cases were rising in the middle of October 2020, ITV news – a popular commercial television channel – broadcasted a roundtable of coronavirus experts. Some argued for a national lockdown to be introduced, while others called for the UK to live alongside the virus. During the debate Devi Sridhar, Chair of Global Public Health at the University of Edinburgh, explained that, "Right now, we are taking a hit to the economy, major economic damage, without the public health benefit, which, in a way, is kind of a worst of all worlds" (Channel 4, 2020).

Sridhar's argument draws our attention to the two main ways of talking about the effects of the pandemic: public health and the economy. But it also emphasizes the way this conversation is often underpinned by numbers, even if they are not explicitly stated. If we rework her statement using data, we can see the implicit logic: 'At the moment, in England, GDP growth is negative and cases and deaths linked to COVID-19 are high, which is the worst of all worlds.' The fact that Sridhar does not need to refer specifically to these indicators demonstrates the way these metrics do not just represent the economy and health during the pandemic – they have come to stand in for these two phenomena. To talk about the economy and health, generally means talking about economic metrics and public health indicators. The numbers have become the phenomena they attempt to represent.

This is not unusual – it has become common to use 'the economy' and 'GDP' interchangeably. But the pandemic witnessed the rapid establishment of two contradictory data bounds that rested on the ability of numbers to stand in for health and the economy. *Trade-Off* was the dominant way of thinking of the pandemic in England – it positioned official measures to combat the virus as improving public health but coming at the cost of the economy. In her quote above, Sridhar was talking from the second, less popular, data bound: *Protect Both*. It emphasized that public health measures could protect both the economy and health, whereas a lack of state intervention would mean both the economy and health would suffer.

This chapter explores both data bounds, examines where they came from, how they were supported by data and how they related to policy. In doing so, it highlights how dominant data bounds can often be a result of political decisions rather than a reality based on the most robust data.

Trade-Off

When Sridhar was interviewed in October 2020, *Trade-Off* dominated England. It used data to reinforce the idea that lockdowns were necessary public health measures but detrimental to the economy. To appreciate how this data bound emerged as the dominant way of thinking about the crisis, we need to consider its history.

When England, Scotland, Wales and Northern Ireland all went into their first national lockdown on 23 March, it was largely in response to the exponential increase in cases of SARS-CoV-2. While testing capacity was fairly low, it was still clear that the number of people infected threatened the ability of the healthcare system to cope. Rising positive cases would lead to large numbers of hospitalizations – not only would this lead to many deaths directly attributable to those hospitalized, but it also threatened to overwhelm the NHS. These rising cases and hospitalizations pushed the government to introduce a national lockdown. This measure was effective in reducing the number of cases, hospitalizations and deaths.

But as the data for GDP emerged in May and June 2020, it was clear that the UK as a whole had paid the price economically. Monthly and quarterly reports produced by the Office of National Statistics (ONS) were the first to provide estimates of the effect of the lockdowns on GDP. The initial calculations released by the ONS (2020f) about quarter one (January, February and March) detailed that the economy had fallen by 2 per cent. Given that the UK restrictions came into full effect on 23 March, however, the most shocking figures emerged from the monthly estimate for April. The report documented that GDP fell by 20.4 per cent in April – the biggest drop in GDP that the UK had ever seen (ONS, 2020e).

It is important to note that these figures were subject to substantially longer delays than public health indicators. The initial report of a 2 per cent decline in quarter one was published on 13 May – around six weeks after the end of the quarter itself. There was a similar delay for the April report – it was released on 12 June. While the negative effect of the lockdown on the economy came as no surprise, the scale of the national lockdown was met with disbelief by journalists, politicians and many within the public.

Looking at the data in June 2020, it was hard to argue against the idea that the national lockdowns had an unprecedented negative effect on the economy but had protected public health. By 12 June, when the April GDP data was released, the UK had a seven-day average of 82.4 coronavirus-related deaths – this had reduced from the peak in April of 969.7 deaths (UK government, 2021).[1] Back of a postage stamp maths says this was an 11-fold decrease in deaths from COVID-19 compared to a 20.6 per cent reduction in GDP.[2] In other words, these drastic public health measures meant health had improved but the economy had tanked. While this became the dominant data bound of the pandemic, it was not the only one. If we turn to the international stage, *Protect Both* emerges from countries in East Asia and Australasia.

Protect Both

While almost all countries initially entered a national lockdown at the start of the pandemic, there was a noticeable difference in the public health strategies taken by countries *after* they ended these lockdowns. Whereas the UK – and other European and North American countries – took a *mitigation* approach to coronavirus (as described above), other countries in East Asia and Australasia took a different tack. These can be grouped into two distinct strategies: *elimination* – the suppression of the SARS-CoV-2 cases to zero (for example, China and New Zealand) – and *containment* – keeping case numbers very low, so they never increase substantially (for example, Vietnam and South Korea).[3]

Across all three strategies there was an initial trade-off during the first lockdowns between public health indicators and economic ones. But in those countries that took an *elimination* or *containment* approach, this balancing act dropped away: the numbers of cases, hospitalization and deaths were kept very low *and* their GDP growth rates began to rebound. In effect, they protected both health *and* economy metrics.

It was within these contexts where *Protect Both* emerged. This centred on the notion that both public health and the economy could be protected during the pandemic. By keeping cases at zero (or very low), strong border controls meant that much of 'normal life' could resume. While this included important social, cultural and personal activities, it also meant people and businesses resumed normal economic activity. The prevailing logic from economists is that effective management of public health reduced uncertainty in the market. Less uncertainty means that people are more likely to spend money rather than save it, keeping the economy turning (Charumilind et al, 2020; Chetty et al, 2020).

Protect Both could also work the other way though. If countries failed to keep case numbers low, which was the case in much of Europe and North America, they would see a negative effect on the economy due to the increased levels of uncertainty and lower resulting levels of spending. In effect they would fail to protect both. We can observe how this logic sits in clear opposition to *Trade-Off* above, but how does data function within this reality?

As we have seen, every country that went into an initial national lockdown saw a negative effect on its GDP. This included countries who took an elimination, containment or mitigation approach. Therefore, the initial data regarding public health and the economy pointed in the direction of *Trade-Off*. But as countries achieved zero COVID or kept their levels sufficiently low, data began to emerge that supported *Protect Both*.

On the same day as Sridhar's interview in October 2020, the *Financial Times* (2020a) published a data visualization that mapped 'fall in GDP' and 'cumulative deaths per million'. They plotted data from several different countries from across Europe, North America, South America, Asia and Australasia. While they did not draw a line of best fit through the scatterplot, there seemed to be a trend: certain countries saw a small reduction in GDP and fewer deaths, while others saw sharp falls in GDP and lots of deaths.

Almost no countries fell into the areas that would be expected for a *Trade-Off* narrative to be true: high numbers of deaths but a thriving economy or low numbers of deaths and a poor economic performance. Only India saw a large fall in the economy and relatively low numbers of deaths. While international comparisons are tricky, the scatterplot represented a growing body of data that supported *Protect Both*. Not only was there no clear distinction between health and the economy, countries following the mitigation approach actually saw GDP reduce the most *and* higher levels of deaths.

This data, and similar analyses performed in September and October, were crucial to the data bound of *Protect Both*. This is the logic from which Sridhar made her argument that England's approach was 'the worst of both worlds'. In her view, England had failed to protect either the economy or public health because of the *mitigation strategy* that it adopted. To understand how these competing data bounds both claimed to rely on data, and both existed in the same space of public discussion despite being a paradox of each other, we can evaluate the data in each more closely.

Short-term vs long-term data

When we compare the data, we can see two clear differences the time-span that each piece of data covers and the comparative nature of the data. Each set of data comes with its advantages and disadvantages when attempting to understand COVID-19. Short-term data provides excellent real-time numbers on how the pandemic is progressing but lacks an overall picture to make broad evaluative judgements of a country's performance economically or in terms of health. On the other hand, international comparisons reduce the accuracy of data but allow for a comparative understanding of specific strategies to tackle coronavirus.

Trade-Off in England focused on short-term rates for both public health indicators and GDP. The three most common metrics for public health were cases, hospitalizations and deaths. These were rarely presented as cumulative totals – that is, the total number of cases during the entire pandemic. Instead, much of the discussion about these three metrics were either daily totals or a seven-day rolling average. The totals were useful – providing a daily count of reported cases, hospitalizations and deaths – but these numbers could be quite misleading. Reporting delays over the weekend would mean the daily counts announced on Sunday and Monday would almost always be lower than Tuesday. Therefore, the seven-day rolling average soon became the standard way of representing and talking about these metrics.[4] This average flattened out the 'noise' in the data caused by these delays and allowed politicians, experts and journalists to identify whether cases, hospitalizations and deaths were increasing, staying the same or decreasing.

Beyond this, the rolling averages could be linked together. News media reports would emphasize that as more people contracted the virus, then a percentage of those people would be hospitalized with coronavirus and a percentage of those people would go on to die from the virus. This meant commentators could provide something of a future picture: if case rates had only just started to increase, you would expect to see hospitalization rates increase next and then deaths. We can see the successional nature of this data most clearly in the third national lockdown in early 2021. When England went into this lockdown, the seven-day average for cases peaked first on 1 January, followed by patients admitted to hospital with COVID-19 on 9 January and then deaths 28 days after a positive test on 19 January (UK government, 2021).[5] In other words, there is a progression. In this example, cases peaked on day 1, hospitalizations on day 9 and deaths on day 19.[6]

A focus on short-term data can also be observed in the economic metrics. Attention was generally paid to monthly reports of economic performance. This meant that much of the discussion about the economy would rely on the percentage increase or decrease of monthly GDP compared to the previous month or the equivalent month in the previous year. To describe the economic cost of the national lockdown, the GDP figures for April 2020 were compared to those for March 2020, whereas the figures for July 2020 were compared to June 2020 to emphasize the economic benefit of emerging from the national lockdown.

The short-term nature of these measures served an important purpose. In terms of public health, case rates became a very good early indicator of the pressure placed on hospitals and the resulting deaths of people with the virus. Therefore, these metrics were essential for disease surveillance and effective management of the virus. Similarly, early understandings of the economic impact from lockdowns were essential for the government to know where and how it needed to intervene economically. But these short-term rates fail to grasp longer-term trends or overall picture.

Yes, we can see that initial lockdowns caused economic damage to all countries. The USA saw GDP shrink by 4.8 per cent (*Financial Times*, 2020c), South Korea's economy contracted by 1.4 per cent (Jaewon, 2020) and Germany's GDP figure reduced by 2.2 per cent (*Financial Times*, 2020b). But in the countries that did not need to introduce further lockdowns because they followed elimination or containment strategies, the longer-term economic growth was markedly different to the UK. And herein lies the difference between *Trade-Off* and *Protect Both* – the former referred to data covering days, weeks and months whereas the latter used numbers covering quarters and years.

In the international comparison by the *Financial Times* cited above, the focus was not on 'what is happening now' but on 'what has happened over the entire pandemic'. In doing so, the visualization relied on longer-term changes in GDP and cumulative deaths per million people. These figures looked to move beyond daily counts and seven-day rolling averages to data that provided a longer-term picture that was framed in terms of months and years. But this longer-term data involved tricky comparisons of different countries' data to each other – something absent from *Trade-Off*.

International comparisons

These comparisons were generally used to identify which countries were adopting the better strategy. This was the approach taken by the *Financial Times'* infographic but also a number of other organizations who conducted similar analyses. *Our World in Data* published a graph on 1 September that plotted confirmed deaths from coronavirus per million against GDP growth of the latest quarter compared to the previous year (Hasell, 2020). An article in *The Conversation* used a similar visualization that also used deaths per million but plotted change in GDP per capita using a similar visualization on 25 November (Smithson, 2020).

Whether it was the *Financial Times*, *Our World in Data* or *The Conversation*, the same pattern emerged. Those following elimination and containment strategies were doing better than those adopting mitigation. Such a comparison was not possible in *Trade-Off* as it did not make international comparisons and relied on data that was too short-term. But with international comparisons, also come a range of problems. While both data bounds rely on GDP and deaths, the *Protect Both* number narrative has to contend with the problems of comparing these figures across countries.

The problems of comparing death tolls

All of these international comparisons used data that counted all the deaths related to coronavirus in each country. On the face of it, this count should be relatively simple. But different definitions of a coronavirus-related death and lack of testing capacity renders the comparison of death tolls a tricky business.

At the start of the pandemic, The World Health Organization (WHO, 2020a) urged countries to adopt the clinical definition of a coronavirus death due to the lack of testing capacity facing most nations. This meant that doctors recorded whether coronavirus was the cause of death for a person on their death certificate. The doctor could base their decision on a positive test for the virus but did not necessarily need to rely on this information. The idea was that if there were not enough tests being conducted, there would be a large number of people dying from coronavirus who had not received a test for it. By expanding the definition to clinical expertise, these cases could be accounted for. This was adopted by some countries, such as Belgium and Canada, but others, such as the UK and Italy, remained reliant on a positive test for the daily counts of deaths.

This split in how countries counted a coronavirus-related death was highlighted by The COVID-19 Health System Response Monitor who analysed the definition of a coronavirus-related death in 30 countries across the world (Karanikolos and McKee, 2020). They found that half (n=15) relied on a clinical diagnosis. Of the remaining 15 countries, 11 relied on positive tests to determine whether there was a COVID-19 death.

Here we can see one of the fundamental problems of international rankings – the data from each country relied on national data collection that was dependent on different definitions of a COVID-19 death.[7] In general, countries that relied on deaths after a positive test recorded lower numbers of deaths per million than countries who adopted the clinical definition. We can see how this panned out in the context of the UK.

In the early stages of the pandemic, the UK government directed their limited testing capacity to hospitals. This meant that any patient who was admitted to hospital with coronavirus symptoms was tested. If the patient had a positive test and later died, they would be counted as a death from COVID-19. This approach meant that large numbers of elderly people who died in care homes did not receive a test in March and April 2020, and so were not included in the figures. Data from the ONS that relied on the clinical definition found that by 24 April 2020 there had been 5,890 deaths involving COVID-19 in care homes since the start of the pandemic (ONS, 2020d).[8] Most of these 5,890 deaths would not have been included in the government's daily count.[9] We can see a similar problem in comparing countries' GDP.

The problems of comparing Gross Domestic Product

The ONS (2020a) provides a good explanation of the three main measures of GDP adopted across the world. The first is the production approach (GDP(P)) – it assesses 'the value of all goods and services *produced* within the economy'. The second concerns expenditure (GDP(E)) – it is 'the sum of all final expenditures within the economy, that is, all the *expenditure* on goods and services, which are not used up or transformed in a productive process'. The final approach refers to income (GDP(I)) – it is the sum of 'all income generated by production activity'. The production, expenditure and income approach *should* all equal the same if each measure was perfectly accurate. Given the difficulties in summing up the entirety of economic production, however, these numbers form estimates. They are brought together into an average to provide the best estimate of GDP – it is this average that forms the headline figure.

Each measure relies on a wide range of different types of data. As the ONS are keen to stress, this type of 'national accounting' should not be considered accounting in the traditional sense of the word – calculating the financial performance of a company by relying on comprehensive financial records. Instead, this national accounting process of arriving at a GDP figure is based on 'statistical surveys, forecasts and models'. In this way, GDP is an inferential statistic, as it relies on a sample of data to make inferences about the whole economy. Business surveys are one of the key pieces of data relied upon. In the UK context, this includes the Monthly and Annual Business Survey, the Construction Survey, the Quarterly Stocks Survey, Living Costs and Food Survey, and so on (ONS, 2019). Putting it somewhat simply, the figure for GDP largely rests on the information provided by a sample of businesses.

While surveys are the main form of data for almost all countries making GDP calculations, not all countries adopt the same method. The UK, as we outlined above, combines all three measures when estimating its GDP. Sweden, on the other hand, only uses a production and expenditure approach – to date, there is no complete calculation of the income side (SCB, 2018) – while China estimates its quarterly GDP using the production method only (National Bureau of Statistics of China, 2020).

Stepping away from these three methods of calculation, some countries have witnessed dramatic changes to their GDP after they included information that was previously excluded. In August 2019, Vietnam announced that it would be making changes to its GDP methodology to bring it in line with international standards (Le, 2019). Prime Minister Nguyen Xuan Phuc explained that Vietnam had not previously included the 'shadow economy' in its calculations – this formally recognized 76,000 businesses that were previously excluded (Thuy, 2019). These revisions were also backdated, meaning that the GDP figures for 2010–17 increased by an average of 25.4 per cent annually (UN, 2019).

These sorts of discrepancies between the different measures to assess GDP and the different data that can be included, lead statisticians to offer words of caution when comparing GDP figures from different countries. In response to a ranking of G7 countries by GDP growth in 2020, the Radio 4 production *More or Less* produced a short podcast in January 2021 about macroeconomic comparisons (*More or Less*, 2021). Simon Briscoe, Director of The Data Analysis Bureau, explained that the UK counted the market value of government services differently from other countries. The standard practice across Europe was to approach the market value of education by counting teachers' wages. The ONS, on the other hand, had been developing a system that measured the amount of teaching. Given that national lockdowns involved teachers' wages staying the same, yet their time teaching was considerably reduced, the UK figure provided a much more negative picture. The output of the government sector fell by 23 per cent in UK, whereas the only countries in Europe that saw a reduction by more than 10 per cent were France and Hungary.

Briscoe estimated that the impact of this different methodology was considerable. He explained that the 20 per cent fall in the second quarter of 2020 would have been reduced to 14 per cent if the ONS took the Spanish approach and to around 16 per cent if they took the Dutch one. Whether these estimates are correct or not (and they were subject to some scrutiny), it seems clear that the ONS provide a more accurate picture that is also worse economically. For Briscoe, these differences meant that the UK was simply not comparable to other countries (*More or Less*, 2021).

Furthermore, when we consider the scatterplot by the *Financial Times*, we can see the potential problems of not adjusting for expected growth rates. It is not fair to compare China with the UK given that the former's annual GDP growth has been consistently higher than the UK's for the past three decades. But how well can other countries, with more similar rates of expected GDP growth, be compared to each other? Some guidance is offered by GDP projections before the pandemic hit (IMF, 2020) but, rather than giving definitive answers it only points towards the complexity of comparing GDP across multiple countries.

This leaves us with two data bounds that essentially operate in opposition to each other and both use different data to underpin their claims. *Trade-Off* is supported by rates that highlight how lockdowns in the UK involve an improvement in cases, hospitalizations and deaths but a decline in GDP. Whereas *Protect Both* focuses on long-term data from a number of different countries to emphasize the way countries either succeed in protecting both health and the economy or fail to do so. Importantly, both of these sets of data are accepted in the scientific community (even if they are flawed) yet they serve two very different purposes. But what about the data that is excluded from both?

What about data outside the data bounds?

While there are differences between *Trade-Off* and *Protect Both*, there is a tacit consensus that GDP is a good measure of the economy and cases, hospitalization and deaths are good measures of public health. But these are not the only measures. We can start with GDP.

Alternatives to Gross Domestic Product

The first criticism levelled at GDP is that it fails to measure macroeconomic performance effectively. It is considered a simple calculation that adds up the total value of everything produced by all the people and companies in the country (whether they are foreign or citizens). But this measure does not account for the depreciation of assets, treats natural resources as free and often fails to account for the informal economy (Fioramonti, 2014, pp 4–5). However, it seems that its simplicity is part of its enduring power, as it allows for international comparisons to be made more easily than more convoluted and country-specific measures (Mügge and Linsi, 2020).

A second criticism encourages us to critically engage with macroeconomic performance. Even with a perfect measurement of this phenomenon, what does the overall size (and growth) of the economy tell us? There is no consensus between the size of the economy and levels of inequality – with many pointing to the way levels of inequality increase as the economy grows (Turnovsky, 2013). Therefore, GDP can be explained by a small number of people getting much richer while a larger number of people see marginal, or even no, improvements in their material wealth.

In light of this, authors advocate for alternatives to macroeconomic indicators. The most radical of these is the Gross National Happiness (GNH) by the Kingdom of Bhutan (Balasubramanian and Cashin, 2019). This measure assesses how well the country is doing by using a measure that falls in line with Buddhist teachings regarding happiness, contentment and self-worth. The GNH is based on nine domains, covering psychological wellbeing through to education. While this measure has also faced its critics,[10] it does point to the way GDP is not the only measure of the economy – or society as a whole (Fioramonti, 2013, pp 93–6).

This also brings us to the suitability of the public health indicators: cases, hospitalizations and deaths. These three measures are subject to less systematic and fierce criticism than GDP. In part, this is because they rely on three distinct measurements of public health – people who contract coronavirus but are not hospitalized, people who are hospitalized and people who die from COVID-19. In doing so, it catches those people who suffered from a mild case of coronavirus, those who experienced more serious consequences and those whose life was taken by the virus. But this does not mean there are no other indicators. One measure that was not used by the *Financial Times*' data visualization, but was used by others when comparing deaths across countries, was *excess all-cause mortality*.

Excess deaths

Whereas the coronavirus-related deaths counted all deaths that could be directly linked to the virus, all-cause excess mortality is not concerned with the cause of death at all. It examines all of the deaths registered within a certain time period (generally a year) and compares this number with a baseline number of deaths that would have been expected to occur in that time period. This difference results in the number of extra people who died than would have been expected.

Because this approach is not concerned with cause of death, it incorporates people who died from coronavirus alongside all the other knock-on effects of COVID-19, such as deaths from delayed cancer operations and deaths from those less likely to go to hospital because of fear they will contract coronavirus. It also includes the longer-term effects of COVID-19, including those who would die from deprivation as a result of public health measures and economic impact. But this measure also works the other way too – it includes certain causes of deaths that might have decreased during the pandemic. A prime example of this is the near non-existence of influenza-related deaths witnessed over the winter of 2020–21 in the UK (PHE (Public Health England), 2020b). On top of this, it can also include events that change mortality but have no connection to the pandemic – for example, the heat wave in the UK in mid-July 2021 seems to have been, at least in part, the reason why death registrations from all causes are above the five-year average.

To account for all these deaths, all-cause excess mortality relies on two key measurements. The first is the total number of people who died during the pandemic. In the context of the UK, and many other European countries, this is a highly accurate measurement – almost all the deaths that occur are recorded. The uncertainty emerges in the second measurement – the estimation of a baseline. To approximate this figure, statisticians ask the question: when we examine the same period in previous years, how many people would we expect to have seen die in the period we are interested in *this* year? And in asking such a question, they split 2020 into two realities: the one we lived – where the highly unusual event of the pandemic occurred and caused a certain number of deaths that we can count – and the one we did not live – where the pandemic did not occur and 2020 was similar to previous years. As you might expect, the reality we did not live is much harder to quantify.

One rather simple but commonly used method of establishing the baseline is to use a five-year average.[11] At its simplest level, a five-year average looks at the data from 2015, 2016, 2017, 2018, 2019 for each week in the year, adds them together and divides by five (the number of years). This was the sort of approach favoured by the analysis from the *Financial Times* during the early stages of the pandemic. They compiled data from 'national statistical agencies for 19 countries for which sufficient information exists to make robust comparisons'. Their analysis began at the point when 'death rates in individual countries climbed above five-year averages' (Burn-Murdoch and Giles, 2020).

This approach, while popular, is considered problematic by statisticians as it ignores different population structures and the ways these change. Using the previous five years' data on mortality assumes that there have been no significant trends in the make-up of the population of that country over time. Most European countries, however, are witnessing a gradual increase in the proportion of older people in their population (European Environmental Agency, 2016). The rate of this change, and how that varies across countries, is not taken into account in the all-cause excess mortality calculations outlined above.[12] But what about alternatives to cases, hospitalizations and deaths altogether?

Beyond cases, hospitalizations and deaths

With the emergence of long COVID, a condition that leaves people with long-lasting effects of the virus, there has been a push to recognize how patients can live with the virus long past the point at which they contract it. By January 2021, Layla Moran – a Liberal Democratic politician – explained that around 300,000 people were now living with the long-term effects of the virus in the UK (BBC News Online, 2021b). In either data bound, this long-term effect of the virus is not captured. And it is this long-term result of the pandemic that also underpins people's call to recognize the mental health damage caused. Young Minds, a mental health charity, pointed to their January 2021 survey to emphasize the scale of the problem. They explained that 67 per cent of people they surveyed thought that the pandemic would have a long-term effect on their mental health (Young Minds, 2021).

When we consider the economic and public health measures that were not included in either *Trade-Off* or *Protect Both*, we can see how both are set within broader conventions regarding GDP, cases, hospitalizations and deaths as appropriate indicators. In excluding certain data, both emphasized what *mattered* by what was *counted*. Within this consensus of what matters, each data bound relied on distinctly different ways of using this data. Whether you think one set of data was more convincing than the other, what is important is to see how both data bounds use data that *is* broadly convincing to politicians, journalists and the public. But how can we account for how *Trade-Off* was more successful in the UK than *Protect Both*? To answer this, we need to appreciate how official policy mutually reinforced the data bound.

How policy structures data bounds

As I explained, the initial lockdown of the UK seemed to establish *Trade-Off*. But the act of locking down a country was common to almost all countries during the initial spread of the virus in early 2020. The important moment for *Trade-Off* was in the way England, Wales and Northern Ireland emerged from the first lockdown adopting a 'mitigation' strategy.[13] This meant that they did not aspire to reduce cases to zero or very low levels. Instead, they took an approach that can be seen as 'living with the virus'. Over the summer months case numbers were relatively low, but in September and October 2020 these public health metrics began to rise. This resulted in a national lockdown in November 2020, followed by the third national lockdown in January 2021.

We can see how this policy choice of 'mitigation' mutually reinforced *Trade-Off*. The first national lockdown was empirical proof that government intervention would reduce cases but also reduce GDP. When cases began to rise again, in autumn 2020, this thinking was repeated: the government must intervene again to protect health, even though it will cost the economy. In fact, this intervention was deemed too much of a 'light-touch' as it failed to stop the spread of the virus and only had a marginal effect on the economy. This made the third national lockdown more stringent than the second, again emphasizing the need to control the spread of the virus, despite the damage it would wreak on the economy. With each decision, *Trade-Off* was established as common sense in the UK.

The power of such a relationship can be identified in the way Sridhar herself, an advocate of *Protect Both*, was resigned to the need for a national lockdown in her interview in October 2020. She explained that "unfortunately I don't see any other way now in England" because they had followed a mitigation strategy. Here we can see the power of this data bound: it legitimized national lockdowns and foreclosed space to imagine beyond the current strategy. This allows us to understand why *Protect Both* failed to get 'noticed' in the UK (Beer, 2016). Yes, it was supported by data. But it ran counter to common-sense understandings about the pandemic in the UK (a common sense that was constantly reinforced by policy decisions).

This relationship between common sense, data and policy shines a light on the nature of data bounds. As the chapter detailed, both sets of data are technically accurate and reliable (with their own technical deficiencies), but they work to underpin paradoxical claims concerning COVID-19 policy. Knowing this encourages us to push against the all-pervasive notion of data as revealing an objective truth (one explored in more detail in Chapter 3).

In the case of this chapter, we need to think of data bounds as something distinctly national. This can be seen most clearly in how data is produced, with different statistical practices for health and economic data across countries (outlined in the 'International comparisons' section above). But data is also localized in the way it becomes meaningful. From the vast banks of data produced by each country, a selection of this quantitative information is emphasized in the construction of data bounds – with policy driving which data is selected and which is not.

Countries following an elimination or containment strategy would point to long-term health and economic trends, often comparing their performance to other countries. Whereas countries taking a mitigation approach often fixated on nationally bound short-term data about rising or falling cases, hospitalizations, deaths or GDP figures. After a period of time, these cemented *Protect Both* and *Trade-Off*. These data bounds structured the way certain countries saw the pandemic itself. In light of this, we need to push against the idea that data 'speaks for itself' as a universal and objective piece of knowledge and towards the way data is 'spoken about' and 'spoken with' in local and subjective contexts. Or, to paraphrase Louikissas (2019), how *all data bounds are local.*

Despite this, the power of data bounds is how they present themselves as universal, objective and absolute. This emerges from 'quantitative realism': the idea that the quantitative provides access to an objective reality, revealing it to those who quantify it. Such a notion can often feel all pervasive. But how is this quantitative realism established? Chapter 3 maps this out in relation to two measurements that circulated during the pandemic.

3

Quantitative Realism Underpins Data Bounds

If it was possible to locate a 'start' to the pandemic in the UK, then most people would point to the end of January 2020. This was when the first two officially confirmed cases of coronavirus were found in North Yorkshire, a county in northern England. *The Guardian* covered the contact tracing effort by Jeremy Hunt – a Conservative politician and chair of the Health and Social Care Select Committee – and Public Health England. They led with the headline 'Hunt begins search for "close contacts" of the two UK coronavirus cases' (Boseley and Walker, 2020).

There is a capital 'p' Political chapter that could be written here. It might emphasize the failures of the UK government to identify the scale of the threat posed by a virus that had forced China to place millions into lockdown measures just over a week before. If this argument seems an unfair one to level at the government at the end of January 2020, then it is one that holds more weight when we consider that widescale government intervention was not enacted until 16 March. But the point about government inaction and ineptitude has been well documented in the popular press, academic literature and investigative journalism (Calvert and Arbuthnott, 2021).

So, this chapter takes a different approach. It focuses on the 'close contacts' element of *The Guardian* headline. It examines international scientific and public health literature up to January 2020. The chapter shows that 'close contact' is defined as being *within two metres of an infected person for 15 minutes or more*. In outlining the history of these two parameters – two metres and 15 minutes – the chapter shows that having such a simple binary of 'more dangerous' and 'less dangerous' sits at odds with the unimaginable complexity of the phenomenon of transmission. Despite this, this one-size-fits-all definition of 'close contact' was consistently used in public health messaging. This speaks to the power of numbers to make the unknown

known, establishing quantitative realism even when reality is evasive (Desrosieres, 2002).

Two metres

When we look in the body of *The Guardian* article cited above, we can see that 'close contact' – while presented as somewhat vague – was tightly defined using numbers: 'The PHE [Public Health England] definition of close contact is being within two metres of the infected person for 15 minutes' (Boseley and Walker, 2020).

The first number concerns the measurement of distance by referring to a person being within two metres of an infected person. The second focuses on measuring time: being within two metres of someone for 15 minutes. Taken together, we have a definition of 'close contact' that relies on a mathematical description of space and time.[1] But where did these two numbers come from? Let's begin with two metres.

When scientific papers on transmission refer to two metres, they generally go as far back as the study of Jennison (1942) in the USA. But the story of two metres stretches further back to the 19th century in the work of Carl Flügge, a German bacteriologist and hygienist. Using culture plates (flatbottomed laboratory containers) in a series of experiments in the 1890s, Flügge was the first to find that droplets produced by an individual's nose and mouth contained microorganisms that could transmit disease. These droplets, Flügge found, were relatively large and fell to the ground rapidly – within a few feet (or 1 yard) from the individual (Duguid, 1946, p 472).

It was this finding that had the biggest influence on subsequent science of droplet transmission.[2] But it was the work of two Americans – William and Mildred Wells – in the 1930s that extended this empirical work into the binary framework used today. On the one hand, there are large droplets that do not evaporate before they reach the ground, leaving stains on culture plates in the vicinity of the individual. On the other, there are small droplets that evaporate before they settle on the ground, leaving 'droplet nuclei' or 'residues/aerosols' suspended in the air for longer periods of time (Randall et al, 2021, p 34).[3]

Despite creating this distinction, much of the emphasis from the 1930s to 1970s remained on large droplet transmission – with airborne transmission only gaining widespread acceptance as a route of transmission from the 1980s onwards (Randall et al, 2021, p 34). This meant that during the middle period of the 20th century the focus remained on 1–2 metres as the important distance to discern risk (Randall et al. 2021), with the emphasis generally being placed on one yard (or three feet) as the distance that most droplets travelled (Qureshi et al, 2020).

Therefore, when SARS emerged in 2003, the consensus for proximity was three feet or one yard. We can see this in a paper by Seto et al (2003) who refer to 'within 0.91 m (3 feet)' as the definition for being exposed to SARS. They used this as a guideline to determine the effectiveness of different precautionary measures against contracting coronavirus, finding that those inside 0.91 m who took precautionary measures should have a lower rate of infection compared to those within 0.91 m who did not take any precautions (or as many).

But a growing body of evidence highlighted that SARS-CoV-1 posed a danger to those beyond one yard. Wong et al (2004) referred to a hospital that had a single infected patient yet a cluster of positive cases among medical students. While many of these students were within one metre of the index patient, 'a few ill students were never within 1 m of the index patient' which 'raises the possibility of transmission by other mechanisms' (Wong et al, 2004). They point to transmission by contaminated formites (for example, bedclothes) but also transmission by 'aerosols over a limited distance' (small droplets) (Wong et al, 2004, pp 274–5).

Moving from hospitals to other contexts, Poutanen and McGeer point to the work of Olsen et al (2003) explaining that 'on a flight carrying a symptomatic person with SARS, 90% of the persons who became ill were seated more than 3 feet away from the index patient' (Poutanen and McGeer, 2004, p 221).[4]

These exceptions pushed the scientific community to accept – to varying extents – the transmission of SARS-CoV-1 over longer distances. In their guidelines for isolation precautions from 2007, Siegel et al (2007, p 28) explain that the 'greatest risk of transmission is to those who have had close contact' but that specific cases and controlled experiments show that 'droplets from patients with these two infections could reach persons located 6 feet or more from their source' (Siegel et al, 2007, p 18).

By the time SARS-CoV-2 emerged in late 2019, this meant that countries adopted a distance that sat within this range:

- **1 metre:** China, Denmark and France
- **1.5 metres:** Australia, Germany and Italy
- **2 yards:** USA
- **2 metres:** Nigeria and the UK

As we can see, however, the official measuring unit of the country dictated the distance they elected: the US chose two yards because it uses inches, feet, yards and miles to calculate distance, whereas the UK went for a distance that was 17 cm longer because they elected to use centimetres, metres and kilometres instead.[5] Nevertheless, all countries sat somewhere between 0.91 m and 2 m (or 1 yard and 2.19 yards).

But opting for these one-size-fits-all rules for distance – whether one yard or two metres – provides an illusion of certainty. The way viruses travel in the air through droplets is much more complex than a binary of *inside X distance* or *outside X distance*. Some push for a spectrum rather than an either/or. In 2003, Varia et al (2003) provided a grade of risk for healthcare workers for their investigation of SARS-CoV-1. This ranged from 'high' risk for those with no protection who are within one metre from the case to 'minimal risk' of 'exposure within 3–10 m of the case'.

Even this graded approach is too rigid though. The work of Lydia Bourouiba in the last five to ten years has found that small and large droplets form parts of 'gas clouds' emitted from a person's nose and mouth. These gas clouds are affected by the actions of the person infected – shouting, yelling, coughing or talking – and the atmosphere into which this gas cloud was emitted. All of these determined how far this gas cloud could travel – with certain droplets travelling six to eight metres from a person sneezing (Bourouiba et al, 2014; Bourouiba, 2016). Reflecting on her work in 2021, Bourouiba (2021, p 552) explains that 'it is not the size of the droplets at emission that determines their range but rather the characteristics of the warm and moist gas cloud that is emitted and carries them forward'.

Therefore, different situations mean different types of gas clouds and that means different ranges of transmission. It is worth thinking about just some of the factors to consider for each setting involving an infected person and a person at risk (Pringle et al, 2020; CDC, 2021):

Infected person:

- If they have symptoms (coughing and sneezing), there is an elevated risk of transmission.
- If they are engaging in activities that involve expelling droplets from their mouth (singing, talking, shouting and so on), there is an elevated risk of transmission.

Environment:

- If a space has more people in it, there is an elevated risk of transmission.
- If a space is poorly ventilated (either by windows being shut or being no mechanical ventilation), there is an elevated risk of transmission.

Person at risk:

- If they are very close to the infected person, there is an increased risk.
- If they wear a face mask, there is a reduced risk of transmission.
- If they are vaccinated, there is a reduced risk of transmission.

The factors listed above (and several others not included for brevity) make the science of transmission unimaginably complex. It involves every single interaction between an infected person and a susceptible person in a specific environment across time. Even with all the data we have about coronavirus in the 21st century, this level of detail cannot be accounted for. Therefore, two metres turns a hugely complex, uncertain and context-based phenomenon into a binary between being at higher risk (inside two metres) and at lower risk (outside two metres). In other words, it provides quantitative realism for a reality that eludes exact quantification.

15 minutes

To try to make this marker of risk more sophisticated, the definition of 'close contact' brings in a time parameter. The argument is that within two metres of an infected person for a few seconds is different from being within two metres of someone for a few hours. This is where the story of '15 minutes' comes into play. But where did 15 minutes come from? Why is it the parameter used? What scientific credibility does 15 minutes have? The origins of 15 minutes are more opaque than two metres.

There is no reference to '15 minutes' in literature on SARS-CoV-1 (emerging in 2003) or other transmissible diseases during this period. The figure surfaced during the 2012 MERS-CoV-1 outbreak. From 2012 to 2014, '15 minutes' was used in the European Centre for Disease Prevention and Control (ECDC) definition and other European based literature (Puzelli et al, 2013; Pavli et al, 2014). It also emerged in the UK's MERS Close Contact Algorithm to define 'close contact' – both in 2014 (PHE, 2014a) and in the updated version from 2019 (PHE, 2019).

So, we can say with some confidence that using 15 minutes as a specific rule in relation to coronaviruses (and seemingly any respiratory disease) is no more than 10 years old. But where does the rationale for 15 minutes come from? It is hard to gauge from the literature itself.[6] The reason for this lack of evidence is best explained by the Centers for Disease Control and Prevention (CDC) in a document from September 2021:[7] 'Data are insufficient to precisely define the duration of time that constitutes a prolonged exposure. Until more is known about transmission risks, it is reasonable to consider an exposure of 15 minutes or more as prolonged' (CDC, 2021).

The CDC clearly state that the 15-minute parameter does not come from a specific scientific paper. Instead, it functions as a 'reasonable' definition to distinguish between a 'prolonged close contact' and *not* a 'prolonged close contact'. This distinction is used much like the two-metre guide, to create two groups of risk: higher risk (those within two metres of someone with coronavirus for more than 15 minutes) and lower risk (those within two metres of someone with coronavirus for less than 15 minutes). This makes a lot of common sense. Spending 15 minutes face-to-face with someone does, logically, mean a higher risk of transmission compared to passing someone in a supermarket in a matter of seconds. But there is no clear logic why this number is 15 instead of 12, 13, 14, 16, 17 or 18.[8]

So, we are left with an arbitrary number of 15 minutes that makes common sense but *cannot* be linked directly to scientific literature. This is combined with a similarly arbitrary number of two metres that *can* be linked to scientific literature. They come together to provide the definition of 'close contact' referred to by *The Guardian* in their article from late January 2020 detailing the first two confirmed cases of coronavirus in the UK.

So, the idea of 'close contact' organizes the incredibly complex phenomenon of transmission into two groups: *close contact* – within two metres for 15 minutes or more; and *not a close contact* – outside two metres or within two metres for less than 15 minutes. Given the flaws in the science (and non-science) of these numbers, why were they used for the definition of 'close contact'?

How 'close contact' structured policy

In part, this mathematical precision of the definition was a public health imperative. In late January 2020, when very little was known about the virus, it made sense for the UK's response to rely on previous definitions for coronaviruses. This one-size-fits-all approach allowed the government to provide clear guidelines within its disease surveillance system. The previous guidance on MERS provided public health workers with a clear process: start with a positive test, find those who were within two metres of those people for 15 minutes or more, contact those people at risk, record specific information about them and put in place certain actions (for example, telling them to isolate).

But the story of the close contact stretches beyond January 2020. As the pandemic progressed, this fixed definition was gradually traded for multiple different definitions. By October 2021, official government advice (UKHSA, 2021) referred to five definitions for 'contact' (the term used instead of 'close contact'). One of the five was the original 'close contact' definition, but it also listed four others:

1. face-to-face contact including being coughed on or having a face-to-face conversation within one metre;

2. if they have travelled in the same vehicle or plane as a person who has tested positive for COVID-19;
3. anyone who lives in the same household as another person who has COVID-19 symptoms or has tested positive for COVID-19;
4. being within one metre for one minute or longer without face-to-face contact.

These additional definitions do not just use precise numbers for space and time, they refer to the specific place *and* the behaviour of individuals. There are references to people being in 'the same vehicle or plane' or 'the same household', as well as emphasizing 'face-to-face contact', 'having a conversation' or 'being coughed on'. In doing so, the approach to a close contact is more sophisticated, addressing the pertinent issue of *context* missed in the strict dual-measurement definition from January 2020.

But there were key elements of the government's disease surveillance system that still relied only on this definition. Most notably, the NHS Track and Trace mobile phone application. Using Bluetooth technology, it alerts an individual when they have been within two metres for 15 minutes or more of a person with a positive test for coronavirus (NHS, 2020).[9] In other words, the app did not account for the setting of the interaction or the behaviour of the individuals. This is largely because these apps relied on Bluetooth technology, which needs to deal in parameters that have mathematical certainty – such as two metres and 15 minutes. But it is also because more granular, specific data – such as whether someone is having a conversation with someone else – crossed the boundary of acceptable surveillance. In this way, the NHS Track and Trace app kept the mathematical definition of 'close contact' alive.

It was also embedded within the way people were told to behave during the pandemic. The communications around 'social distancing' centred on being two metres or more away from other people.[10] This was reflected in the markers on the floor outside shops on the high street, signs on the walls of community centres and the way people talked of being a certain distance from someone else. Even when social distancing was officially brought to an end in the summer of 2021, the idea of being two metres from other people remained as a sort of cultural notion of risk.

The original idea of 'close contact', therefore, is still central to ideas of transmission – albeit with additional definitions that look to place interactions in context. But even these expanded definitions rely on the idea that coronavirus transmits through large droplets that travel a short distance from the infected person and drop to the floor relatively quickly. The science of 'gas clouds', however, emphasizes the way droplets travel much further and stay in the air for much longer. And this is supported by a number of studies identifying cases where the virus had transmitted between people who were within two metres of each other for less than 15 minutes (Brogan, 2021; Mack et al, 2021). So, how can we explain the enduring power of both of these numbers?

How numbers organize the unorganizable

To begin, let's return to the two-metre rule. As it was shown above, Flügge's work is contested: scientists argue that droplets *did* travel further than the three feet documented by Flügge but he missed them, so they were not measured. But the actual act by Flügge of measuring three feet is not challenged. None of the research says 'actually Flügge's tape measure was not accurate when we measured "a few feet" in the late 1890s'. This allows those refuting his claims to make their own measurements, compare their measurements to Flügge's and, when they differ, challenge his research.

Therefore, the context in which scientific measurements took place are treated as irrelevant. The 'three feet' measurement made by Flügge in late 19th-century Germany can be referred to by Bourouiba investigating gas clouds in 21st century USA. Bourouiba can then demonstrate that the gas clouds can travel up to 'six to eight metres' and this measurement can be referred to by future work. The numbers may contradict each other (six to eight metres is much further than three feet), but the actual measurement stays the same. This makes numerical measurements timeless and universal: they can be produced in specific times and places, travel to other times and places, be reproduced, discussed and refuted and then travel to other times and places.

This use of numbers sits in direct opposition to subjective notions of space and time. Numerical descriptions are not interested in whether something *feels* like a long time or something is *experienced* as a short distance, it looks to say exactly and precisely how much time has passed or distance travelled. The way numbers have come to occupy this privileged position in human knowledge is not a simple story, but the final part of this chapter points to the importance of language, measurement and documentation. In doing so, the chapter draws heavily from Porter's (1995) book *Trust in Numbers* and the writing of Latour (1986) in 'Visualization and cognition'.

Numbers are part of the language of mathematics (Porter, 1995). Woodin sees numbers (1, 2, 3, and so on) as representing nouns, and operational signs (for example, +, –, /, × and =) serving as verbs. These two components are used together according to a set of standardized rules, developed over thousands of years (Woodin, 1995). In this way, a formula, such as '3 + 3 = 6', can be seen as a correct mathematical sentence, whereas another, such as '3 + 3 = 4', is an incorrect one. This can be compared to the way one sentence in the English language, such as 'the cat wore the hat', is correct but 'wore cat the hat the' is not. But the mathematical language has specific characteristics that set it apart from the English language.

English language is developing constantly, varies with cultural contexts and sits firmly within ambiguity, whereas mathematics is a very rigid and unambiguous language – it describes lengths, counts and masses rather than emotion, love and experience. Sure, researchers try to make numbers account for these things by quantifying the social world, but these projects, when relying on numbers alone, always fall short. The very reason they fall short is why mathematics can speak with such authority about an objective material reality.

It exists to precisely define 'things' in our world. Whether you are in Tokyo or Kinshasa, the mathematical language is constant: three apples remains three apples, 40 cm remains 40 cm and 15 kg remains 15 kg (even if some countries elect to use pounds instead of kilograms or inches instead of centimetres). The universality of measurements is a product of the universality of the rules of mathematics. For Porter (1995), this means the people who use mathematics afford the numbers they create a 'mechanical objectivity'. That is, the numbers are trusted because they are the product of a process where the subjectivity of the researcher is stripped away by the strict, universally accepted rules of mathematics – leaving an objective piece of quantitative knowledge.

The way mathematical measurements are seen as timeless and ever-present is not entirely due to linguistics though. There is a mutually reinforcing relationship between the language of mathematics and the devices used to measure space and time. To measure '15 minutes' means relying on some sort of clock – an old-fashioned analogue version, a newer digital type or the extremely accurate atomic clock. Whatever the exact device, they are all governed by a certain mathematical organization of time: there are 24 hours in a day, 60 minutes in an hour, 60 seconds in a minute, 100 milliseconds in a second, and so on. This means that these mechanical devices *materialize* the linguistics of mathematics.

In the case of time, the clock is the physical manifestation of the idea that mathematics can describe a fixed notion of time. But the clock is not just the product of the language of numbers, it also serves to reinforce mathematics itself. The clock demonstrates the very idea that time can be organized into units (days, hours, minutes, seconds, milliseconds, and so on). It would be a much harder concept to grasp if all clocks ran fast or slow and we had no way to distinguish which one was which. If you asked someone the time and they said it was somewhere between 1:30 pm and 5:14 pm, you would have very little use for a clock. And a scientist would have even less use for this imprecise measurement device.

The same can be said for devices measuring distance, such as those used in the 'two metre' literature. They rely on a certain unit of measurement, whether it be metric or imperial, to function. But the very idea of a standardized distance relies on these measurement devices being consistently reliable. Imagine if all tape measures could only give a wide range of possible lengths, meaning that measuring the length of a piece of string gave a result of '56 to 341 cm'? These measurement devices would be far less useful, and they would also undermine the very idea that length can be expressed precisely as a number. Therefore, it is this mutual relationship between technology (the device measuring) and the language of mathematics that sustains the idea that time and space can be measured according to a set of units (for example, seconds or centimetres) wherever someone is on earth and at whatever period of time.[11]

This allows scientists to perform all sorts of measurements. In the late 1800s, Flügge could measure the distance between a person and a set of culture plates – resulting in the 'few feet' statement. It also allowed Bourouiba et al (2014) to use high-speed cameras to measure the 'six to eight metres' distance that a gas cloud travelled from a person's mouth. In this process, the researchers do not just measure space though. They *document* the numerical measurement. Whether this is done in a lab book or on a computer, whether entered manually or automatically captured, turning these measurements into some form of document is crucial for how these numerical measurements become fixed (Latour, 1986).

When these measurements are documented, it is rare that researchers will deal entirely with numbers and operational signs. They generally combine mathematics with two other languages. The first is the national language – for the purposes of this chapter, it will be referred to as English. This could be when researchers provide some context to measurements – this can include the description of the unit (for example, 'four *metres*'), the title of tables (for example, 'Lengths of distance travelled by gas cloud on 24 September 2013') or for notes, clarifications and reminders (for example, this plant looks undernourished, might be an anomaly, keep an eye on it). The second language is geometry: lines, curves and shapes that allow the researcher to organize findings in tables, where each row and column are separated by a straight line. Or the plotting of results on graphs that need curved lines for the findings and straight *x*- and *y*-axes.

It is hard to overstate the importance of the work that these documents do. They turn the complexity, mistakes, confusion and messiness of often very long processes of research into a scientific text that presents this research as ordered, precise and logical. These initial documents are then used to create formal public scientific texts. Perhaps the most iconic of these documents is the academic journal article – from which the story of the two-metre parameter is largely derived.[12] But this is complemented by a number of other important texts, from conference proceedings to pre-print articles.

These texts then circulate in print and online in academic circles. For most authors, their document's circulation is confined to universities, think tanks or government policy. They are read by other academics in the field, discussed at conferences and referenced in other scientific research. Some authors, however, see their work break through into public discourse. This can occur for a number of reasons, for instance: the work is 'open access' – where the public do not need to pay to access the document, it is communicated through institutional PR channels (for example, a press release by a university), the author is well known in media circles, the piece of work is particularly newsworthy, and so on. Whatever the route, these texts are almost always the foundation: it is from these documents that quotations are picked out, where conclusions are referred to and where specific noteworthy findings are drawn from.

The process of documentation does an important piece of work. Latour (1986, pp 11–14) argues that these documents turn scientific knowledge into something immutable and mobile. That is, they can travel across the world as unchangeable pieces of knowledge. This allows for 'distant or foreign places and times' to be 'gathered into one place in a form that allows all the places and times to be presented at once' (Latour, 1986, p 29). Here we can see an overlap with how numbers function. Part of the way numbers can be ever-present and timeless, as described above, is because the scientific documents they exist within are immutable and mobile.

Binding together the sciences

Therefore, measurement, linguistics and documentation combine to give numbers their peculiar power of rendering unknowable things known. The language of mathematics is about precision and objectivity, measurements are a materialization of this precision and objectivity, and documents take numerical measurements, combine them with other forms of representation, and create immutable and mobile scientific documents that reinforce the universality and timelessness of mathematical measurements.

The power of these three interlinking phenomena (measurement, linguistics and documentation) is starkly highlighted when we consider the shaky evidential basis for either 'two metres' or '15 minutes'. For these two cases at least, we can see how numbers are often used in science as 'markers of certainty' for knowledge that lacks any notion of certainty itself (Van Witsen, 2019). This quantitative realism binds together science – as a set of knowledge, organizations and individuals.

The certainty of quantitative realism is the glue that sticks together the way we think of science, the vast array of organizations dedicated to scientific research and thinking, and the individuals who *do* science. The power of numbers to provide this certainty predates the pandemic. But, in the story of 'two metres' and '15 minutes', we can see how quickly, effectively and totally numbers can turn the unknown into a piece of scientific knowledge. A task that can be seen as all the more necessary for a scientific community weighed down by a lack of data and incomplete explanations about the pandemic.

It seems that for the two cases in this chapter, the sciences did achieve this goal of certainty. The definition of close contact circulated as numerical facts across news organizations, social media, political communication, public health literature, private messenger chats and so on. They served to render the incredibly complex world of transmission into a neat binary of either being below or above the numerical threshold. This influenced the way people managed the risk of coronavirus in their day-to-day life, the parameters of models looking to predict the spread of the virus, the way that health practitioners attempted to reduce transmission and how organizations mitigated the risk of transmission in their establishments. While much of this effort was necessary, the absolutism attached to these two figures was misguided – and perhaps the lack of communication about their uncertainty played some role in the way the public understood and acted out these measures.

Looking outside of the pandemic-specific context, we can also ponder on the implications this chapter has for the theories and theorists that frame it. To explain how numbers organize the unorganizable, the chapter relied on two pieces of work: Porter's (1995) book *Trust in Numbers* and the writing of Latour (1986) in 'Visualization and cognition'. Both are considered foundational, but far from 'contemporary'. Neither frames their discussions within notions of 'the digital'. Therefore, what do the contemporary case studies of 'two metres' and '15 minutes' have to say about the nature of linguistics, measurement and documentation in the 21st century?

There has been little change in the *language* of mathematics – at least at its most basic level. We have seen the rise of digital *measurement* devices, namely digital scales, digital clocks and laser distance measurers. These have enabled scientists to make increasingly accurate, reliable and precise measurements of time and space – further cementing the idea that the language of mathematics can be materialized in the world around us. But it is the process of documentation that has undergone the most landmark shift.

As Latour emphasizes, scientific documents had the power to be both immutable and mobile in the pre-digital era. In the 21st century, journal articles, pre-print articles, books and conference proceedings can be (near) perfectly replicated, sit in domains that can be accessed from anywhere across the world instantaneously and then shared through complex networks of social media, private media and news media. Their improved replication is combined with the ability of publishers to change texts too – errors in articles can be corrected and amendments can be added. So, scientific *documentation* in the digital era allows for the hyper-mobility of texts that are both more perfectly replicable and easier to change.

And this means the numbers sitting in these documents are also afforded this digital immutability and mobility. Figures, statistics and indicators circulate across news media, social media and private communications at breakneck speeds; rearing their heads simultaneously and instantaneously in different cultural and national contexts. When combined with the universal linguistics of mathematics and the way measurement devices materialize this notion of numbers, quantitative realism has been granted even more power in the 21st century.

The shoring up of quantitative realism has strengthened data bounds. The more the world feels quantifiable and becomes quantitative, the more data bounds are entrenched within society. In this way, we can see the belief in quantitative realism materialize in the existence of data bounds. But this faith in the quantitative does not just emerge from the language, measurement and documentation of the sciences. It is also an abstract – non-mathematical – phenomenon. Chapter 4 outlines how large numbers create a sense of 'hugeness' that underpins our understanding of what data can say about the world around us. In doing so, it further develops our understanding of how the linguistic meets the mathematical.

4

Quantitative Realism Is Mathematical and Abstract

The story of 'one billion items of PPE' begins at an English government press conference on Saturday 18 April 2020. These daily briefings, streamed live across radio and television broadcasters, comprised of one elected politician from the Conservative Party and one or more scientific advisers to the government. These officials provided an update on the latest data, the policies being introduced and implemented, and the longer-term strategy to deal with the pandemic.

At this particular press conference in mid-April, Dan Hewitt from the popular television broadcaster ITV asks three questions to Steven Powys, the National Medical Director for National Health Service (NHS) England, and Robert Jenrick, Secretary of State for Housing, Communities and Local Government. It is worth outlining each in full:

1. How have we found ourselves in the situation where we are dangerously low on PPE?
2. Why hasn't the government had a plan B here, getting small, medium, large manufacturers to produce PPE?
3. Do you accept the worries of NHS doctors and nurses that we have spoken to today that by downgrading your PPE guidance, by not providing proper PPE, you are putting their lives at greater risk? (BBC News, 2020a)

Robert Jenrick, after a brief stutter, emphasizes how the public are in awe of social care and NHS staff and how nobody wants to see these people worried about whether they will have the correct equipment. He emphasizes how the government has to do more to get the PPE to the frontline but also stresses how it is a huge challenge given the global demand for equipment. Despite this, he explains that the government is making progress: "Today I can report that a very large consignment of PPE is due to arrive in the

UK tomorrow from Turkey … which will include 400,000 gowns" (BBC News, 2020a).

This statement starts a strange ball rolling in government rhetoric around supply and distribution of PPE – one that bounced along from Saturday to Sunday to Monday to Tuesday and, finally coming to a stop, on Wednesday. It is important to provide a day-by-day play of how this panned out.

The day following Jenrick's comments, Michael Gove – a government minister – told Sophie Ridge from Sky News that the 400,000 gowns from Turkey were delayed (Walker, 2020). In the evening press conference on Sunday, Gavin Williamson (Education Secretary) confirmed that the shipment had been delayed until the following day. He faced seven questions from journalists, five of which concerned PPE (BBC News, 2020b).

Responding to the first question by Hugh Pym from the BBC, Williamson explained that there had been a huge effort over the past few weeks that had led to "a billion items of PPE" being brought into the country. It felt like whatever Robert Jenrick could do – with his claims of a whopping 400,000 gowns – could be bettered by Williamson, who introduced the fabled claim of 'one billion items of PPE'. The shift from numbers of items in particular shipments to the total number of items across all shipments took root in government communication over the next few days.

On the Monday, Oliver Dowden (another government minister) was challenged on Radio 4's morning show about a further delay in the shipment of 400,000 gowns from Turkey. In response, he reiterated the claim made by Williamson that the government had acquired one billion items of PPE since the start of the pandemic. In the press conference that evening, Rishi Sunak (Chancellor of the Exchequer) responded to Hugh Pym's question about PPE supplies by restating that people on the frontline deserve the equipment they need to do their jobs. He restated the claim made by Williamson and Dowden, pointing to the 'billion pieces of PPE' that has already been delivered (BBC News, 2020c).

In the daily press conference the following day, Matt Hancock (The Health Secretary) explained that "since the start of the crisis, we have now delivered over one billion items of PPE" (BBC News, 2020d). On 22 April, it was the turn of Dominic Raab (the Secretary of State). In the weekly debate in parliament called Prime Minister Questions (PMQs), his retort to the opposition leader's questioning on PPE was the now stock line of 'one billion items'. He went on to repeat the figure that evening in the daily press conference (BBC News, 2020e). After this press conference on 22 April, four days after Robert Jenrick talked about 400,000 gowns from Turkey, the number of one billion largely disappears from political rhetoric.[1]

The use of 'one billion items of PPE' provides a remarkably self-contained moment of statistical rhetoric, one that is instructive of how these types of numbers are used in political communication. There are several aspects to unpack. Considering production, what does the *one billion* items of PPE comprise? At the level of linguistics-meets-mathematics, what exactly does one billion mean? Focusing on political communication, what role does this number perform rhetorically?

Counting cleaning wipes

Before we consider the issue of what one billion actually means in a linguistic sense, it is worth outlining what one billion items of PPE are at a technical level. While government officials did not disclose the composition of the one billion items of PPE, a BBC *Panorama* (2020b) documentary from 27 April outlined what was counted in this total.[2] Leaked data showed that just over half were individual surgical gloves (520,981,860) and around 14 per cent were flimsy plastic aprons (144,113,400). Also included in the total were items not traditionally associated with PPE, including cleaning products, wipes and waste bags.

Conventional definitions of PPE centre on equipment that provides a protective layer between a person and a hazard. The inclusion of wipes and waste bags, for example, runs counter to the list of different items of PPE included in the World Health Organization's (2020b) guidance from 19 March 2020. Wipes and waste bags, while important components of a health and safety, are *not* items of PPE. Their inclusion should be treated as nothing more than statistical fabrication.

Other elements of this spreadsheet do refer to PPE but involve an inventive way of conceiving of these items. The counting of individual surgical gloves, for example, runs counter to the way gloves in general are purchased, worn and stored as a pair, and also sits at odds with how these gloves would actually be used by healthcare professionals – it would be hard to imagine a nurse working with one hand inside a glove and another not.

These counting procedures should be considered all the more bizarre when we revisit the comments of Jenny Harries, Deputy Chief Medical Officer to the government, during the Gavin Williamson press conference from 19 April 2020. In response to a question about the supply of adequate PPE, Harries is clearly annoyed at the over-focus on the shipment of gowns from Turkey. She calls for a grown-up conversation that thinks "carefully about what has been achieved" by the government rather than "lumping all the PPE together" because PPE is "not a homogenous mix at all". Sadly, the 'one billion items of PPE' figure cited by Williamson moments later not only homogenizes PPE but throws in millions of extra items, too (BBC News, 2020b).

But timing is everything; this *Panorama* documentary was aired five days after the short lifespan of our statistic ended. While their findings were useful at the time as fact-checking-meets-investigative-journalism, they made relatively little difference to the way one billion items of PPE functioned within political discourse at the time. So, we should move forward with the knowledge that what counted as PPE was farcical, but centre the conversation on how this number functioned as a piece of communication. To do so, we have to appreciate the way numbers are a language in themselves.

Numbers as language

The way numbers underpin quantitative realism, as outlined in Chapter 3, obscures the relatively recent history of our number system. From around 1200 to 1500 AD in Europe, there was a gradual shift away from Roman numerals towards Arabic numerals. In Crossley's (2013) examination of 1,398 manuscripts created between 1200 and 1500, he found that only 7 per cent of manuscripts in the 13th century had the 'new' numbers, compared with 17 per cent in the 14th century and 47 per cent in the 15th century. It is hard to overstate the significance of this shift in mathematical language in Europe.

Let us begin with some mathematical syntax. What makes Roman numerals so different from Arabic numerals? First, we have to deal with what is called 'place value notation' – the way figures, placed in a certain order, represent an individual number. Take Arabic numerals, the mathematical language we use today. If we want to express 'four thousand, three hundred and two' as digits, we construct the following '4302'. Here we have four digits, so we know that the last number represents the number of 1s we have, the second to last the number of 10s, the third to last the number of 100s and the fourth to last the number of 1000s. In other words, 4,302 can be thought of as four separate numbers that are added together.

$4 \times 1000 = 4{,}000$
$3 \times 100 = 300$
$0 \times 10 = 0$
$2 \times 1 = 2$

This is what we commonly refer to as the base ten counting system. With this system, you only need ten digits (0, 1, 2, 3, 4, 5, 6, 7, 8, 9) and you can construct any whole number you like. The Roman numeral system, as used in Medieval Europe, did not have a system as simple as this. First of all, it did not express numbers as single digits. Even for small numbers, several different symbols needed to be used in the right order.

1 = I
2 = II

3 = III
4 = IV
5 = V

While slightly inconvenient when dealing with small numbers, the use of these symbols for large numbers presented a considerable challenge. More often than not, a dot was placed between numbers to designate a place value system. As Menninger (1969, pp 284–5) explains in his iconic book, the number 1,485 would be expressed as 'M.IIIIC.IIIIXX.V' (1000.400.80.5). The process became more complicated if there was a need to express 'zero', because Roman numerals had no symbol for zero itself. So, scholars would use the word 'nula' to represent 'nothing' for numbers such as 1,405.[3]

Furthermore, this system could only reach 3,999 because within the logic of Roman numerals, you could only put symbols that were the same next to each other a maximum of three times. Given that M (1,000) was the largest number in Roman numerals, the highest number in this system could not exceed 3,999. Past this point, a grouping system was adopted. This used the same symbols but placed a horizontal line (a vinculum) on top of these symbols to signify that it should be multiplied by 1,000. For example, 1,550,000 would be written as $\overline{M}\ \overline{D}\ \overline{L}$. Broken down into its individual parts, $\overline{M}$ is a 1,000 × 1,000 (1 million), $\overline{D}$ is 500 × 1,000 (500,000) and $\overline{L}$ is 50 × 1,000 (50,000).

Given the complication of the Roman numeral system, and its somewhat clunky way of expressing large numbers, it is no surprise that the Arabic numeral system was adopted across Europe. This new number system was not just good at expressing larger numbers more simply, it also easily allowed for complex calculations. Those using Roman numerals in Medieval Europe would use counting boards – flat boards inscribed with horizontal lines that represented different denominations (for example, five or ten or 50). Counters were then used to keep track of how many different denominations there were (for example, three counters on denomination five would mean a total of 15). These 'counter boards' as they were called in Medieval Europe allowed for subtraction and addition relatively easily but were poor at conducting multiplication or division (Oliver, 1997).

Arabic numerals meant these counter boards were replaced with paper and inscription tools – a form of representation that did not just allow for the multiplying and dividing of numbers but for complex calculations involving tens of thousands and millions. Furthermore, because there was a written record of calculations using Arabic numerals, these calculations could also be checked at a later date.

The importance of this shift for Europe cannot be overstated. Many point to this new mathematical language as the foundational logic to modern accountancy. The emergence of double-entry bookkeeping in the early 16th century was largely attributed to a new number system that allowed for the representation of a lot of numbers in a small space and dealt in negatives as well as positives. This type of accountancy is positioned by many as influential in deriving the very notion of 'capital' as being something you could increase or decrease (Dean et al, 2016).

Turning towards science, we can consider the way mathematics became *the* language of the sciences from the 16th century onwards. Not only did this change the nature of science itself – a process that many call the *mathematization of science* – but it also introduced mathematical concepts that were foundational to entire modern disciplines (Jongsma, 2001). The development of calculus – the mathematical concept of continuous change – would have been close to impossible with Roman numerals (Westfall, 1980). And without calculus, mechanical, structural, aerospace and civil engineering would be unrecognizable (not to mention contemporary mathematical modelling).

The meaning of big numbers

For our purposes, however, we can see the contribution of Arabic numerals in a much more confined sense: the way the new system allowed for the mathematical expansion of scale. In other words, the rise of big numbers. The largest number expressed using Roman numerals would be in the low millions – the $\bar{\text{M}}$ (1000 × 1000) being the highest single symbol – whereas Arabic numerals led to the emergence of the billion, the trillion and the quadrillion in the 15th century. It was in this century that the French mathematician, Nicolas Chuquet (1882), famously explained 'million, the second mark byllion, the third mark tryllion, the fourth quadrillion, the fifth quyillion, the sixth sixlion, the seventh septyllion, the eighth ottyllion, the ninth nonyllion and so on with others as far as you wish to go'.

The ability of Arabic numerals to go 'as far as you wish to go' marks a distinct shift in European episteme from a limited and clunky representation of huge numbers to a relatively easy increase of scale by adding a successive number of zeros to the end. Such an easy representational change – from 1,000,000 to 10,000,000, for example – can obscure the scale of the increase with each additional zero. It can often feel like a shift from a million to a billion is more of a doubling than a thousand-fold increase in size. This was captured well in the UK General Election of 2019 by Ash Sarkar (2019), a political commentator, when she explained the difference between millionaires and billionaires: "If you got £1 every 10 seconds, you'd be a millionaire within 4 months. If you got £1 every 10 seconds, you'd have to wait 310 years before you hit your first billion."

Here, Sarkar captures what is at stake with the Arabic numeral conception of numbers. Yes, it becomes very easy to arrive at large numbers (and perform calculations with them) but it is also easier to lose a sense of the scale of these large numbers. A comic strip style drawing posted on Twitter does a good job at visualizing this confusion (XKCD Comics, 2018). It highlights how 100 million is often perceived to be larger than one billion because of the '100' placed before million being larger than the 'one' before billion. This is despite the fact that the latter is ten times larger than the former.

Up to this point, we have discussed numbers in terms of mathematical grammar. But as Sarkar's example points out, we generally engage with numbers as adjectives. That is, numbers describe the amount of something, whether that is one billion items of PPE or one British pound sterling.

In the English language, numbers operate as adjectives that are completely divorced from the noun being described. Whether it is three elephants or three years, the number does not change form. There are examples where a single word combines the number and the noun – for example, bipedal or tricycle – but these are mainly combinations of the number (bi- and tri-) with a pre-existing noun (pedal and cycle). Much rarer are words that cannot be readily parsed into two separate words. Think of duet, duel and trial and then try to imagine two 'ets', two 'els' or three 'als'. This is all to say that a number occupies a very privileged position within the English language to describe the quantity of any noun (Menninger, 1969).

When we are using numbers past a certain scale, it is hard to actually comprehend or understand the noun being described. Such a conundrum is not a contemporary phenomenon either. In the 16th century, Descartes wrote about the issue of trying to imagine a thousand-sided shape (chiliagon) in his Sixth Meditation (Descartes, 1996). In the early 20th century, Albert Camus wrote the following in *The Plague* about imagining millions of dead people.

> He tried to put together in his mind what he knew about the disease. Figures drifted through his head and he thought that the thirty or so great plagues recorded in history had caused nearly a hundred million deaths. But what are a hundred million deaths? When one has fought a war, one hardly knows anymore what a dead person is. And if a dead man has no significance unless one has seen him dead, a hundred million bodies spread through history are just a mist drifting through the imagination. (Camus, 1970, p 31)

Work from psychology seems to provide empirical evidence that numbers, past a certain point, become less intuitive and more abstract in nature. The work of Dehaene (2011) provides an excellent argument for humans (and other animals) possessing a sense for relatively small numbers (1, 2, 3, 4, 5) that becomes progressively worse as the numbers increase. As this abstraction increases, he explains that Arabic numerals become more useful as representations of size.

The problems of scale also surface in other ways. Research from York University (Jenkins et al, 2018) about facial recognition found that the average person in their research could recall 5,000 faces. While this number feels quite high – it is still far below the types of numbers people are meant to comprehend in public discourse, for example, hundreds of thousands, millions, billions and so on. Our relatively limited ability to store and recall individuals is attributed to the biological evolution of humans living in relatively small communities. In fact, this is the origin of 'The Dunbar Number': the estimated number of people that one individual can maintain stable social relations with is said to be around 150 (Dunbar, 2010).[4] Whether evolutionary or not, the take-home message is that humans struggle to comprehend large numbers of things as unique, individual things. Instead, we conceive of the things being amalgamated in terms of an abstract notion of scale that is expressed as a specific number (say, 1,348,256).

So, we can perhaps see how large numbers are better understood as abstractions than scientific encapsulations of exact scale. This can be noted in the way numbers are used as hyperbole. In their analysis of a five million-word corpus of conversations recorded across the UK and Ireland, McCarthy and Carter (2004, p 179) found that dozens, zillions, millions, hundreds and thousands were often used to exaggerate for effect or emphasis. We can note here a distinction between actual numbers that are adopted for hyperbole – for example, dozens, millions, hundreds and thousands – and numbers that are not mathematical terms but operate exclusively as a rhetorical device – a zillion. What is remarkable is the way these real numbers and non-real numbers operate in a similar fashion.

In this way, big numbers seem to represent 'hugeness' rather than an exact number (Espeland and Stevens, 2008; Billig, 2021). In doing so, we combine the thrust of mathematics to describe large quantities of things and incorporate the abstract nature of these large numbers.

Such a conception of numbers has a longer history in human thought than the more rigid scientific notion: the Egyptian symbol for 100,000 is a tadpole, representing the countless numbers in the mud of the Nile when the water retreated every year; the Indian name for the lac louse, an animal that would appear in countless numbers, comes from the Sanskrit number *lakh* (100,000); while the Greek word for 10,000 was *myrios* (countless) and probably emerged from the Greek *myrmex* for ant (Menninger, 1969, p 122). It is this older concept of numbers that emphasizes the way quantitative realism is about mathematics *and* an abstracted sense of 'hugeness' and 'bigness'.

Adopting this older concept is not to reject the ability of mathematics to accurately capture the large number of items. Rather, it is to emphasize the importance of embracing the abstract nature of large numbers into meaningful notions of 'hugeness' or 'bigness' – where people can feel that numbers can capture large quantities *as well as* see how they do so through mathematical syntax. From this position, we can then appreciate how this mathematical-abstract notion of numbers can wield power as a rhetorical device.

One billion items as political rhetoric

This section approaches the 'hugeness' of 'one billion items of PPE' as functioning within political communication on three levels. In doing so, it draws inspiration from, and makes a vital contribution to, the field of statistical rhetoric (Roeh and Feldman, 1984; van Dijk, 2000; Avilés, 2016; Assa, 2019; Lawson and Lovatt, 2020; Billig, 2021).

Unprecedented crisis met with an unimaginable number

If we are to take one billion items of PPE as a gesture of the 'hugeness' of PPE supplies and delivery, we need to examine how it operated in relation to the 'unprecedented' nature of the crisis. Our concern here is not whether the crisis was unprecedented or not, such a discussion is for others to have, but how such a concept of 'unprecedented' became an important political tool.

The notion of the 'unprecedented' emerged before the first articulation of one billion items of PPE. It was used by government officials in describing both the crisis itself and the response to the crisis - both were positioned as unprecedented. A day after Boris Johnson advised against all unnecessary contact in March, he explained that such steps were 'unprecedented since World War 2' (UK government, 2020b). Five days later, he referred to the economic support provided by the UK as 'unprecedented steps to prop up businesses, support businesses and support our economy' (UK government, 2020e). This trope was firmly rooted within political rhetoric – the unprecedented crisis of COVID-19 was being met by an unprecedented response from the government. Such a narrative underpinned the way PPE supply and distribution was characterized.

In his opening statement on 21 April, Matt Hancock explained that 'protective equipment delivery is an operation of unprecedented scale and complexity' (BBC News, 2020d). We can observe a coalescing of the unprecedented demand for PPE to protect against coronavirus and the unprecedented supply that has met that demand. In other words, the unmatched response is meeting the unforeseeable crisis of PPE.

> 'Enormous effort, and international effort.' (Williamson in BBC News, 2020b)

> 'Every resource of government has been deployed.' (Williamson in BBC News, 2020b)
>
> 'We are doing absolutely everything we can, and straining everything we can, to get the equipment they need.' (Sunak in BBC News, 2020c)
>
> 'We are working day and night to expand that supply base.' (Hancock in BBC News, 2020d)

Here we can observe two key notions of size. First, we have the hyperbole of size, expressed as 'enormous' (extract 1). This articulation, however, was less common than the second notion: the totalizing notion of scale, where 'everything' is being done. All the time, all the money, all the energy and all the resources are geared towards PPE procurement and distribution. It is into this wider discourse of size that 'one billion items of PPE' is articulated. But what does this process do to the very thing being quantified?

The itemization of Personal Protective Equipment

Let us begin by taking the government's calculation of one billion at face value. In other words, let us imagine each individual item counted is in fact an actual item of PPE (for example, an apron or gloves). What is immediately apparent is that it is important to have access to such large numbers of PPE during a pandemic. This is an obvious, but important, point to make. But that is all this number can tell us – that the government has had in its possession, since the start of the pandemic, one billion individual items of PPE that have come from a manufacturer and are to be sent to those who are perceived to need it. Let us consider this notion further.

The one billion items of PPE serves to *itemize* PPE. It depicts PPE as something that exists as separate items stored in a warehouse on individual pallets that are wrapped in cling film. In doing so, it removes PPE from the complex processes of multiple supply lines, chaotic delivery schedules, distribution to staff at hospitals, the way it is worn by individuals and how it is disposed of. These processes involve a range of infrastructure – the ferries, aeroplanes and lorries to deliver PPE and the warehouses, hospitals, care homes, doctors' surgeries and hospices that store it – as well as people, including the warehouse operatives, NHS Trust managers, delivery coordinators, nurses, doctors, care home staff, and so on.

By stripping this context from PPE, and bringing *all* the items of PPE together into a mass of one billion, these processes are erased. While the erasure of production and supply lines is clear, we can also observe how PPE is divorced from its very purpose: to protect people from coronavirus. When we bring back into the fold the technical details of the statistic – and recognize how cleaning wipes, among other things, were included in the total – the cleavage between the reality of PPE and the 'one billion items of PPE' grows wider. In essence, this statistical rhetoric strips away the 'personal' and 'protective' from PPE and leaves us with equipment. Or, in the language of the press conferences, we are left with one billion 'items'.

It is this notion of PPE that seems to underline the frustrations of journalists at these press conferences. The scale of individual items of PPE acquired by the government is clear – one billion – but what does this actually mean for those on the front line who need equipment? Laura Kuenssberg, The Political Editor of the BBC, articulates this mood well in her question to Matt Hancock on 21 April:

> 'Day after day we are hearing from people on the frontline that they do not have enough. How do you explain the gap between what you are saying about getting everything in place and what people who are working on wards or working in care homes actually experiencing?' (BBC News, 2020d)

This discrepancy between the power of the billion to emphasize unprecedented action by the government and the reality of PPE supply and usage on the frontline touches on different aspects of PPE that this figure sought to erase.

The four Personal Protective Equipment problems the figure tried to erase

We can think through four well-documented problems with PPE that the one billion statistic attempted to push against rhetorically: lack of adequate stockpiles, inadequate procurement and distribution networks, prioritizing the NHS over social care and the changing guidance for PPE use.[5]

Lack of adequate stockpiles

The premise of one billion items of PPE is that the pre-existing stockpiles are inadequate to deal with the current crisis. Certainly, COVID-19 provides a considerable challenge in the protection of frontline staff, but should we expect the government to have considerable stockpiles already? If you only listened to the government press conferences, you would be forgiven for thinking that the government had done everything possible. In the press conference from 19 April, Jenny Harries answers the first question from the journalists by emphasizing that the UK was one of the best prepared countries in the world (BBC News, 2020b). This is partially true. The problem was that the UK government was relatively well prepared for the *wrong type* of pandemic.

On 3 August 2020, the Chartered Institute of Public Finance & Accountancy (CIPFA) published an 86-page report titled *How fit were public services for coronavirus?* In this document they explained that the government stockpile of PPE 'was designed for a flu pandemic and did not contain gowns or visors'. These gowns and visors would have proved vital in protecting against coronavirus, which 'can be spread easily by coughing and breathing' (Davies et al, 2020, p 18). This is despite the outbreaks of SARS and MERS – two notable coronaviruses – in 2003 and 2014, respectively. Not only was PPE the victim of a focus on the wrong pandemic but it also seems to have fallen prey to the ideology of austerity between 2010 and 2020.

The report documents that the government led by Theresa May (2016–19) 'chose not to procure visors or safety glasses for all staff due to concerns about the high cost' (Davies et al, 2020, p 18). This was despite the call from the expert advisory committee New and Emerging Respiratory Virus Threats Advisory Group (NERVTAG) recommending that gowns should be added to the stockpile of equipment in 2019 (BBC Panorama, 2020a). In fact, *The Guardian* documented that the value of the UK stockpile of equipment fell by 40 per cent between 2013 and 2019 (Davies et al, 2020).[6] It seems that the government needed to acquire one billion items of PPE because there was a large deficit of necessary equipment.

Inadequate production and procurement networks

Faced with an inadequate stockpile of PPE, the UK government and the NHS raced to secure supply of PPE from across the globe. This presented problems because of the lack of resilience in the UK's supply chains. There was a lack of capacity within the UK to produce PPE, so any domestic supply had to be set up on an ad hoc basis with private companies that volunteered to switch their production to PPE. Given the lack of domestic supply, a bulk of the PPE was supplied from abroad. This presented problems too.

The CIPFA (Davies et al, 2020, p 24) report explained that before the pandemic, the UK 'has primarily relied on getting PPE through "just-in-time" supply chains from suppliers in East Asia'. A *Financial Times* report highlighted how NHS Supply Chain – a government body that procures goods for the NHS – 'had no experience of directly sourcing PPE overseas, and was accustomed to securing it through UK-based intermediaries' (Foster and Neville, 2020).

In light of this, local NHS trusts often 'took procurement into their own hands, only for NHS England to subsequently restrict them from procuring certain goods' (Davies et al, 2020, p 24). The result of this was that the one billion items of PPE, and any future procurement, were subject to the volatility of international demand – the UK would consistently compete with other countries to secure the PPE they needed. But it was not just the stockpiles and procurement of PPE that the one billion statistic looked to erase, it was the way this PPE was distributed.

Prioritizing the National Health Service over social care

Mirroring a wider process of favouring the NHS over the social care setting (namely care homes), hospitals were prioritized over other frontline settings. In fact, it is worth reflecting on a piece of advice published by the government on 25 February 2020 called *Guidance for social or community care and residential settings on* COVID-19. The extract, deeply worrying to those aware of what happened during March and April 2020 in care homes, should be read in full below.

> During normal day-to-day activities facemasks do not provide protection from respiratory viruses, such as COVID-19 and do not need to be worn by staff in any of these settings. Facemasks are only recommended to be worn by infected individuals when advised by a healthcare worker, to reduce the risk of transmitting the infection to other people. It remains very unlikely that people receiving care in a care home or the community will become infected. (UK government, 2020d)

While this advice was retracted on 13 March 2020, it highlights two remarkable things about the government's position during this period: first, COVID-19 was considered to be something that would occur in hospitals; and second, care home staff do not need to wear PPE. The government did start to provide PPE to care homes around 16 March, but the items were inadequate for staff. This meant that adult social care institutions often had to procure items themselves in a market characterized by soaring demand and inflated prices (Davies et al, 2020, p 38). This was compounded by the economic backdrop for those in social care. Whereas the NHS is funded by the taxpayer, nursing homes are 'means-tested and mostly run by private companies partly funded by local authorities that have suffered years of budgetary cutbacks' (Clarke and Plimmer, 2020).

This discrepancy between social care and the NHS was part of a wider problem in the government's response to coronavirus. Not only was the public health message centred around protecting the NHS, but testing and PPE efforts were also geared towards NHS staff. While such a focus is necessary when a new virus might overwhelm hospitals, it also ignored other aspects of healthcare: primary care (for example, GPs) and social care. The significance of such a focus was largely deflected at the time by politicians but was recognized later (20 May) by Robert Buckland, the Secretary of State for Justice, in an interview on Sky News (McGuinness, 2020).[7]

The imbalance between planning for the NHS and social care emerged in public discourse with the rise of 'unaccounted' deaths occurring outside hospitals. The daily figures released by NHS England and the Department for Health and Social Care (DHSC), and communicated in the daily press conferences, only referred to deaths in hospitals of those with a confirmed case of coronavirus. These did not include those dying outside hospital settings – care homes, prisons, hospices and at home. Of these settings, care homes were by far the worst affected. Between 2 March and 12 June 2020 (the first wave), the ONS reported that there were 19,394 deaths that involved COVID-19 in care homes (ONS, 2020c).[8] Without the provision of face masks, as well as other responses such as testing, this total is much higher than it could have been (HSJ, 2020). So we can see how the rhetoric of hugeness conceals a vital aspect of PPE – it must be distributed to the right people at the right time for it to be effective. In the case of care homes, this distribution clearly failed.

Changing what classes as suitable Personal Protective Equipment

The explanation regarding care homes outlined above is not to say that PPE for NHS staff was seamless either. They faced consistent shortages in supply and the worry that they would be left unprotected in the face of coronavirus. But there was a further issue that seems to have been the catalyst for the one billion items narrative that emerged in the press conference on 19 April. The day before, Dan Hewitt from ITV asked three questions (all of which are introduced at the beginning of the chapter). The most important one in this context was: "Do you accept the worries of NHS doctors and nurses that we have spoken to today that by downgrading your PPE guidance, by not providing proper PPE, you are putting their lives at greater risk?" (BBC News Online, 2020a). This question emerged from Public Health England's (PHE) decision to change their guidance for using PPE. The document, published on the Friday before the press conference (17 April), explains that 'some compromise in process is needed to optimise the supply of PPE in times of extreme shortages and we have agreed that the use as outlined in this document is appropriate within health and safety legislation and provides appropriate protection for health and care workers' (PHE, 2020a, p 1).

The change of best practice due to supply shortages raised alarms for those in the news media. A BBC article drew attention to the fact that staff were being asked to reuse certain equipment where possible. The article quotes the report directly, explaining how staff are to reuse '(washable) surgical gowns or coveralls or similar suitable clothing (for example, long-sleeved laboratory coat, long-sleeved patient gown or industrial coverall) with a disposable plastic apron for AGPs (aerosol-generating procedures) and high-risk settings with forearm washing once gown or coverall is removed' (BBC News Online, 2020). While PHE were keen to stress that the new guidance was legal, it raised concerns about healthcare workers being expected to change their use of PPE due to shortages rather than scientific evidence.[9]

The power of huge

By telling the story of one billion items of PPE, we can see how this figure is akin to Billig's (2021) idea of 'semi-magical round numbers' – where curiously rounded numbers have a certain magical property in political rhetoric.[10] But its 'magic' goes beyond being rounded off to the nearest billion, it was the 'hugeness' of this statistic that made it peculiarly powerful.

To understand 'hugeness' means seeing quantitative realism as rooted in the mathematical logics of scale, but rejecting a hard take that can *only* see these figures as mathematical. Instead, these big numbers are semi-magical in their scale – a scale very few actually understand but one that many can relate to. This abstracted notion of numbers combines with the mathematical to underpin quantitative realism and allows 'one billion items of PPE' to serve a specific political purpose: allowing the Conservative politicians to emphasize the scale of the response to the coronavirus, while simultaneously obscuring the multiple failures in said response.

But the academic contribution of this chapter should not overshadow the gravity of the case study itself. The notion of hugeness in political rhetoric can serve a multitude of political purposes – from referring to a technically accurate figure about millions in poverty that supports a progressive politics or to strategic use of a misleadingly large number to deflect valid criticism of government practices. If this chapter was a case about the former, hugeness could be hopeful. Sadly, the 'one billion items of PPE' was more of an example of the latter. It used numbers to prop up a reality characterized by misinformation, fakery and sleight of hand.

By expanding the category of PPE so wide, the number becomes statistically meaningless. This fabrication is then used to cover up a series of government failures, ones that can be laid squarely at the door of the incumbent government. Successive Conservative governments failed to stock the right PPE for this pandemic because of poor planning and the adherence to the policy of austerity, but also failed to recognize the importance of COVID-19 in social care settings during the early stages of the pandemic. All of this culminated in the forced changing of how PPE should be used, meaning staff had to reuse single-use items. Deaths as a result of the lack of PPE – of both staff and patients – can never be accurately quantified, but the introduction of mandatory face mask wearing in certain indoor spaces – for example, shops and supermarkets – from 24 July 2020 to 19 July 2021 emphasizes the importance of PPE in reducing transmission. Using a large number to deflect from this reality is more about political opportunism than the effective management of a public health crisis.

The next chapter examines this relationship between quantitative realism and data bounds more closely. Instead of emphasizing how quantitative realism is believed because of the power of language, measurements and documentation (Chapter 2) or through the mathematical-meets-the-abstract (this chapter), it emphasizes how the need for data bounds drives quantitative realism itself.

5

Desire for Data Bounds Underpins Quantitative Realism

On 27 May, WIRED UK online (Temperton, 2020) reported that *The Plandemic* conspiracy documentary had been 'viewed more than eight million times across Facebook, Instagram, Twitter and YouTube'. Just over three weeks later, on 18 June, the UK section of Yahoo! News released an article titled 'COVID conspiracies: 7% of Britons think there is no hard evidence that coronavirus exists, poll suggests' (Wells, 2020a). Both statistics attempted to capture the public's belief in misinformation.

One number looked to tie belief with views: eight million people had viewed *The Plandemic* – a documentary that argued SARS-CoV-2 was deliberately released to control the world's population. The other figure was based on a traditional approach to public opinion: a set of survey questions asking people about their belief in misinformation. These two statistics were given their life through the narrative of the 'infodemic'. This portmanteau of information and pandemic is described in most detail by the World Health Organization (WHO, 2020c):

> An infodemic is an overabundance of information, both online and offline. It includes deliberate attempts to disseminate wrong information to undermine the public health response and advance alternative agendas of groups or individuals. Mis- and disinformation can be harmful to people's physical and mental health; increase stigmatization; threaten precious health gains; and lead to poor observance of public health measures, thus reducing their effectiveness and endangering countries' ability to stop the pandemic.

It is at this point in the chapter that I face a juncture. I could take the section above as contextual – the foundation upon which a piece of research would examine why people believe in misinformation, where people receive disinformation from or how researchers can prevent people from believing in fake news. This is the path taken by much of the research on the infodemic – see the work of Brennen et al (2020) and Roozenbeek et al (2020) for two noteworthy examples from early in the pandemic. The approach by this chapter, however, follows a different tack. To do so, we must begin with the premises of the infodemic itself.

Rethinking the infodemic

If we take the infodemic – as defined by the WHO above – at face value, we must recognize its premises. First, that misinformation circulates online. Second, that this misinformation leads to mass non-conformist behaviour. This chapter accepts the first premise – it has been empirically demonstrated that misinformation does exist online (Brennen et al, 2020; Nielsen et al, 2020). The problem comes with the second premise to the infodemic – what this chapter calls 'mis-behaviour' (a play on the word misinformation and misbehaving).

'Mis-behaviour' involves people believing false information they are exposed to online and offline, and continuing to act in the same way as they did before the pandemic. This sits in direct opposition to informed and responsible behaviour – where people follow official information about the pandemic by increasing the distance between themselves and other people and reducing the amount of people they meet. The fear for those putting forward the infodemic is that 'mis-behaviour' would significantly derail government plans to tackle the pandemic. But just how much of this type of behaviour actually occurred during the first wave of the pandemic (the period from which our two cases for this chapter come)?

Where was the 'mis-behaviour'?

During the first lockdown in the UK, which came into force on 23 March 2020, the UK government provided data at the daily press conferences on how well the public were adhering to lockdown rules. They referred to movement data – across motor vehicle, bus and train transport – to highlight a substantial drop in the level of movement after the announcement of the national lockdown (UK government, 2020b). In the words of Professor Jonathan Vam Tam, the data showed 'that the British public are following the advice given to them' (UK government, 2020c). This was complemented by survey data from early April 2020 that highlighted how people had substantially reduced their number of individual contacts in any given day (later published in BMC medicine as Jarvis et al, 2020).

As the lockdown wore on, both sets of data consistently pointed towards the UK public following the official guidelines. This was also supported by data from the private sector. Google's mobility reports – using the location history setting – showed that substantially fewer people were travelling to work, to shop and engage in leisure activities and to transit hubs, such as bus stations (Ritchie, 2020).

The data from the press conference, and the way it was interpreted by politicians, academics and journalists all pointed towards the same conclusion: the public were largely complying with the rules. In fact, this level of compliance was beyond what scientists expected. The Imperial College London epidemiological model – widely referred to as the evidence that forced the government to implement a national lockdown in late March 2020 – assumed 70 per cent of people would comply with self-isolation for seven days if they had symptoms, 50 per cent of households would quarantine if a member of the household had symptoms, and 75 per cent of those over 70 years of age would social distance (ICL, 2020).

The high adherence to lockdown rules meant that the government's decision to enter a national lockdown on 23 March 2020 was successful in bringing case numbers, hospitalizations and deaths down. It seems that the misinformation that was circulating online in the early part of the pandemic never materialized into widespread 'mis-behaviour' that reduced the efficacy of public health measures.[1]

Therefore, if data was discussed within the idea of 'the infodemic' it should have served to emphasize its *lack* of threat. Nevertheless, this data bound persisted throughout the summer of 2020. Why was this the case? The argument put forward by this chapter is that the infodemic became a vital component in the way the news media reported on the pandemic.

Trust us: we are not misinformation

If the infodemic is about misinformation, the news media is about information; if the infodemic is about nefarious producers of fake news, the news media is about responsible journalists; if the infodemic is about preying on a gullible public, the news media is about informing and encouraging criticality (Carlson, 2018, pp 379–86).[2] In many ways, this is classic identity formation – using something that represents the opposite of your practices, morals and politics to define yourself.[3]

We can see how this process of identity formation is articulated in the WIRED UK's coverage of the eight million views of *The Plandemic* video. There are the nefarious producers of this conspiracy video – the article refers to the 'star of the Plandemic conspiracy theory video' as the disgraced scientist Judy Mikovits and positions the video within the broader 'anti-vaccination' movement. The article then describes how this production was shared across multiple platforms (Facebook, Instagram, Twitter and YouTube) as it 'escaped the conspiracy theory "echo chamber" and exposed scores of people to dangerous, scientifically-baseless views'. The piece went on to argue that this 'exposure' presented a real risk that the 'response to the greatest health crisis in a generation' would be undermined 'with potentially deadly consequences' (Temperton, 2020).

The narrative above, while convincing, lacks a certain oomph. There is no notion of scale attached to the ramifications of misinformation, so it is hard to know how important such a threat is to the handling of the pandemic. This is where numbers come in. More specifically, this is where the two numbers from this chapter come in:

> 7% of people believe there is no hard evidence that coronavirus exists. (Wells, 2020a)

> [*The Plandemic* video has been] viewed more than eight million times across Facebook, Instagram, Twitter and YouTube. (Temperton, 2020)

The threat of misinformation is encapsulated above in two figures – the 7 per cent of Britons who believe in misinformation and the eight million people across the world who viewed (and probably believed) *The Plandemic*. We can see this most clearly when we trace the life of these figures.

The life of both figures

Let us begin with the eight million views statistic. On 9 May, during the first week of *The Plandemic* being online, *The New York Times* explained that the video had been 'viewed more than eight million times in the past week' (Alba, 2020). The fact that this number surfaces in the US should come as no surprise given the distinctly US nature of *The Plandemic* – produced by an American, starring an American scientist and a narrative that centres on The Chief Medical Advisor to the President of the United States.

This number circulated within North American news media discourses for the next two weeks before it was first used by *The Telegraph* (Temperton, 2020) – a right-wing broadsheet news outlet based in the UK – on 26 May. A day after, it was covered by WIRED UK. That is not to say that *The Plandemic* video was not being covered in the UK news media before this – it was just being reported without specific reference to any number of views.[4] Across both the North American and UK coverage, the number generally functioned in a relatively benign fashion by pointing to the scale of the crisis:

> In just over a week, the trailer racked up more than 8 million views across YouTube, Facebook, Twitter and Instagram. (Brewster, 2020)

> In that same time period, the video was viewed more than eight million times across Facebook, Instagram, Twitter and YouTube. (Temperton, 2020)

Some articles, however, directly interpreted the figure as meaning something more than a broad notion of scale. In the following quote, we can see how the figure is used to say something about 'our human nature' because the video carries something that 'people want to hear': '"A lot of what (Mikovits) is saying, people want to hear," says Nora Benavidez, the director of U.S. Free Expression Programs for PEN America. "And they love it. Eight million views mean that it's playing into something in our human nature"' (Rivas and Jones, 2020).

Whereas the eight million figure had distinct stages to its life that were spread across two national contexts, the 7 per cent figure was relatively self-contained to two days of media coverage. This was driven by the two organizations that produced the research: Ipsos Mori, a market research company, and King's College London, a UK university. Their respective press releases, both published on 18 June, included the finding that 7 per cent of people believe there is no hard evidence that coronavirus exists (Beaver, 2020). It was this figure that was picked up by the news media. Yahoo! News (Wells, 2020a), AoL (Wells, 2020b) and MSN (2020) all used an identical headline: 'COVID conspiracies: 7% of Britons think there is no hard evidence that coronavirus exists, poll suggests.'

The mirroring of each other's headlines was reflected in the content too – each news report was pretty much identical to the other. This reflects the contemporary digital news ecosystem, where content circulates (in remarkably similar form) across multiple news sites (Davies, 2009). But there was also significance in the 7 per cent figure too – an importance that pushed the headline to be repeated by multiple news organizations.

So, we have one number that has a more drawn out, transnational existence, and another that received a lot of attention in a short amount of time. But there is also a difference in what the specific type of numbers *do* within this data bound. They both emphasize scale, but in different ways. The eight million figure works to emphasize the 'hugeness' of the infodemic by drawing on the abstracted nature of multiple millions. The percentage of 7 per cent does a different type of job – it emphasizes the existence of a marginalized, yet still threatening, group of people who believe in misinformation.

But how much can these survey findings and digital data tell us about the scale of the infodemic? What do these two ways of measuring the public actually say? Should we use them to base our opinions about the infodemic? In answering these questions, we can interrogate the quantitative realism underpinning the infodemic.

Surveys and 7 per cent

The use of surveys to understand 'the public' can be placed within a broader shift in the late 18th and early 19th century towards statistics.[5] In

Western Europe, nation-states were developing proto-bureaucratic systems that collected increasingly larger amounts of information on their citizens, culminating in comprehensive censuses. To understand this information, they worked in partnership with the emerging field of quantitative researchers called statisticians (Desrosières, 1998, p 10). This partnership was geared specifically towards statistically identifying social problems and providing state-led reforms to remedy them (Porter, 2003, p 238). In fact, the work of early statistics was almost inseparable from the nation-state. This is evident in the word itself: statistic is derived from the German *statistika*, which itself is derived from a combination of Italian and New Latin terms for 'the state'.

As states adopted statistics to govern their citizens, more and more of the social world was quantified. The work of Adolphe Quetelet (1796–1874) typified the expansion of statistics during the middle of the 19th century. Working alongside the French state, he argued that perceived random individual action – such as crime, marriage and suicides – could be understood as regular (and therefore predictable) when one amalgamated all of these individual actions (Desrosières, 1998, p 10). The differences at an individual level were erased at the level of 'society' as variation gave way to what Quetelet called *l'homme moyen* (the average man). He went on to link these social averages (of crime, marriage, suicides and so on) to causal factors, such as 'nutrition, schooling, religion, and laws' (Porter, 2003, p 241). Towards the end of the 19th century, this quantitative logic was being applied to psychology, forming a new discipline of psychometrics that attempted to measure certain qualities of the brain (Michell, 2003, 2021). In this way, social behaviours, and the things that caused them, were increasingly framed within statistical laws of regularity (Lazarsfeld, 1961, p 297).

It is within this context of the 19th century that the story of 'surveys' emerged as important and valid ways of understanding how people would vote in elections. The US election of 1824 saw a number of 'straw polls' commissioned by newspapers who were determined to understand how the public were going to vote. These polls used a variety of different methods to arrive at their prediction, mainly relying on the counting of votes at meetings before the election. These meetings were both regular events that were not specifically about the upcoming election, and events specifically organized for the counting of people's intentions (Smith, 1990).

As these polls did not attempt to position their sample (for example, people at the meeting) within the broader demographics of voters (for example, all the people in the State), these early assessments of public opinion were often wrong in their prediction. It was the Gallup Poll, conducted in 1932, that first introduced the idea of randomly selecting a group of people to represent the whole population in question. This more sophisticated approach to polling was realized in the correct prediction of the 1932 Presidential election – one that saw Franklin D. Roosevelt (FDR) win a landslide. Many point to this moment as the turning point of polling data. This new method meant the poll was catapulted from a method that sometimes arrived at the right result to a robust piece of knowledge that made accurate predictions (Squire, 1988). The success of the new approach to polling went hand-in-hand with the rise of surveys more widely within academia, most notably in the social sciences. By the 1950s, the survey was seen as a credible way to predict election results *and* understand 'society' more broadly (Mills, 1959).

Of course, there is a much longer and more convoluted history of surveys, polls and quantification of the social world.[6] But it is enough here to say that survey data became the main way that public opinion was represented and discussed during the mid to late 20th century across the news media, academia and politics. It was this method that was used by Ipsos Mori and King's College London to arrive at the claim that '7% of people believe there is no hard evidence that coronavirus exists' (Duffy and Allington, 2020).[7]

The survey was completed between 20 and 22 May 2020 by 2,254 UK residents aged 16–75. To decide who should be included in the survey, the researchers used stratified sampling. This approach rejects total random choice of participants and favours selecting people randomly from categories so they are more representative of the whole population under study. To do so, the researchers set quotas for different age groups within gender, religion, working status, social grade and education that reflected the make-up of the UK population (based on data from the ONS). Then participants were randomly selected until that quota was met. In effect, this approach tried to make the 2,254 people who took part representative of the whole UK population. Once this sample was selected, they relied on Ipsos Mori to conduct the surveys online.

This survey asked three types of questions: conspiracy beliefs, health-protective behaviours and sources of information. At a basic level, this allowed for the researchers to provide *descriptive statistics* regarding specific questions. They would ask the computer software to count how many responses to a question were 'true', 'false' or 'don't know'. This would be presented either as three separate counts (for example, 1,000 people said true) or as a percentage (40 per cent of people said true). It is through this process of *descriptive statistics* that our figure emerges.

Within the section on 'conspiracy belief', a statement read: 'There is no hard evidence that coronavirus really exists' (Duffy and Allington, 2020). The respondents had to select either true, false or don't know. According to the Ipsos Mori report, 7 per cent said 'true', 11 per cent said 'don't know' and 82 per cent said 'false'. Let us consider this particular finding in more detail. From this single question – with three options – those conducting the survey put out their claim that 7 per cent thought that there was no hard evidence that coronavirus exists. This runs counter to more complex surveys that approach the issue of belief in conspiracies through a large set of implicit and explicit questions (often including Likert scales instead of categories of answers). The problem of such a simple approach can be understood through three hypothetical survey participants.[8]

First, what about Simone who has heard there is no evidence that proves COVID-19 exists but is fairly sceptical of the claim? They do not completely believe in this statement or completely disbelieve in this statement either, but they also have enough of an opinion to not feel comfortable clicking 'Don't know'. They think that whatever option they choose will not really reflect what they think. So, they decide not to choose 'Don't know' because they do not like being unsure. They flip a coin for true and false. It lands on heads, so they click 'True'.

Simone encourages us to appreciate the incredibly rigid format of surveys: a standardized online tool that requires participants to answer pre-selected questions by selecting from a limited number of answers. While this standardization is somewhat necessary for quantitative analysis, it comes at a considerable cost. People often select an option that does not reflect what they would have answered if they were afforded more flexibility. It is not to say this does not happen in more open-ended research methods – for example, interviews – but that surveys are *the* most rigid of all.

Second, what about Muhammed who has never come across this conspiracy theory before? They come across this question and consider their response. They draw on their previous experience of COVID-19 and their critical judgement, while simultaneously second-guessing the intentions and expectations of the survey (and researchers themselves). In the end, they consider that the use of 'hard evidence' must distinguish this question from 'normal evidence'. They think that there is 'normal evidence' but think that all of the uncertainty around the data, the origins and the treatment of coronavirus must mean that scientists do not have any 'hard evidence' to speak of. So, Muhammed clicks 'True'.

Muhammed really taps into the assumption that participants arrive at surveys with their ideas about the topic fully formed and ready to be articulated. Such an approach rests on an empirical naivety that still abounds in much of the hard social sciences. This ignores the co-constructive, ambiguous and negotiated nature of surveys as captured in Muhammed's experience. Rather than arriving with answers to survey questions already formulated, the participants often construct their opinion in dialogue with the survey itself.

Third, what about Dave who has been on furlough for six weeks in their studio apartment during the national lockdown? They arrive at the survey having spent the last month and a half mainly confined to their small home in the city centre. They cannot meet up with friends, they cannot work and they can only leave the house once a day. The focal point of their frustration is levelled squarely at scientists, politicians and journalists for restricting what they can do.

They are notified that they have been included in a survey about coronavirus and arrive at the third question in the 'beliefs' section. They are pretty certain there is hard evidence for COVID-19 but they really wish there wasn't. So, a part of them thinks that if they select 'false', then maybe coronavirus might turn out to be a hoax so they can start working, seeing their friends and leaving their increasingly cramped flat. Dave clicks 'True' and moves onto the next question.

Dave's experience speaks to the highly contingent nature of surveys – they capture a person at a particular point in time and present them with a question that they may answer differently in a month's (or even an hour's) time. At this point in time, our fictitious respondent elects for a confused and conflicting approach to their answer. One month on, and Dave has started work again, they can see their friends and family and they have travelled out to the countryside. On their way home, Dave checks Yahoo! News, sees an article titled '7% of people believe there is no hard evidence that coronavirus exists' and wonders why people would believe such crap.

Simone, Muhammed and Dave encourage us to think about what survey statistics actually mean. They draw our attention to the incredibly rigid format of surveys, the way participants form their opinion in relation to surveys and how surveys capture a person at a particular moment in time. All of which is not to say surveys are useless – far from it. Instead, it emphasizes the need for careful consideration of how to overcome the rigidity of surveys to arrive at meaningful findings. But this is not the only way to quantify the public; the use of digital data has become increasingly common – the method used to arrive at the eight million figure.

Digital data and eight million

This shift towards digital data can be set within a relatively recent history that begins at the turn of the 1990s with the rising use of the internet. The proliferation of digital technologies in the 21st century meant that a growing number of organizations collected, retained and used a wealth of digital data that recorded every action and interaction online (McFarland et al, 2016, p 15). In contemporary research, the idea that this digital data can say something more interesting, more personal and more comprehensive than surveys has taken root. This has not come with the death of surveys – rather they are seen to access different aspects of public life.

Surveys involve the identification of participants through predefined categories of identity (for example, gender, age, ethnicity, religion, and so on) and their views and opinions on particular topics, themes or ideas. Whereas digital data is about 'what people do, their interactions, transactions, performance activities and movements' (Ruppert, 2012, p 119). This shift represents an important movement from *asking questions* through a survey to establish something about 'the public', to recording what 'the public' *does online*.

The 'doing of things' online by the public has also meant the scale of information has expanded considerably – recognized most clearly in the term 'big data'. The widespread collection of data from people's activities online has increased the data about activities that existed beforehand – for example, the quantification of running through apps – but also given rise to new types of information (Till, 2014; Ajana, 2017; Lupton, 2019). Private messaging services, social media and email have created new forms of text and image-based communication that are often linked to internet-specific forms of commercialism: online expressions are collected, aggregated and used for advertising products to specific cohorts of potential customers.

With such a wealth of data, the methodological approaches have also changed. Researchers do not begin by trying to select a random sample of people from which they can make inferences about the total population. Instead, they start with lots of data that has already been collected by other organizations. In this shift from sampling to using all the available data, we can see a distinction between the classic statistician who deals with uncertainty and probability and the data scientist who provides totalizing accounts using programming language (Shepherd and Hearne, 2019).[9] It is within this contemporary shift to digital data that the eight million views statistic sits.

To arrive at the eight million figure, it is most likely that *The New York Times* added up the total number of views the video has received across YouTube, Facebook, Twitter and other video-hosting websites. What is not clear, however, is the exact methodology they adopted for this process. When they referred to other digital data about *The Plandemic*, such as the interactions with Mikovits' account, they referred to certain data analytics software (for example, Crowdtangle and Zignal Labs) (Alba, 2020). The eight million figure, however, was not accompanied with any such methodological explanation.

Therefore, we can assume the statistic was produced through a simple counting process of views across known sites that were platforming the video. Such a process is as simple as the illustrative account below:

- **YouTube:** 3 million
- **Facebook:** 1.5 million
- **Instagram:** 2.5 million
- **Twitter:** 1 million
- **Total: 8 million**

As documented in the previous section, most of the coverage of this number approached the amount of views as representing the scale of misinformation.

But there was one notable exception. The article by *The Telegraph* provides an instructive explanation of what 'views' can mean.

> There are two ways to benefit from watching Plandemic. One is to learn how these documentaries work, and see how they exert an influence on you. The other is to be Judy Mikovits. At one point, Willis asks Mikovits a question that, as usual, seems reasonable until you realise that there's never a follow-up. It's the one you'd ask yourself: 'Why come forward now?' Her reply: 'Because if we don't stop this now … we'll be killed by this agenda.' (Revely-Calder, 2020)

Here, the writer makes a distinct break from the simplistic explanations provided by previous reports. This reflexive account emphasizes the way that 'views' are not necessarily a set of die-hard conspiracy believers. Rather, it can also include the interested observer that is concerned with the nature of conspiracy theory productions themselves. In other words, the journalist themselves might make up part of the eight million views they report on. This line of thought pushes us down a more critical path regarding digital data; one that can be extended through a personal anecdote.

During the early stages of the pandemic, I noticed more and more graffiti referring to a 'plandemic' – a conspiracy theory that broadly claims that the pandemic was, in fact, a planned experiment (the purpose of which varies depending on the exact theory). Struck by its similarity to the 'Flat Earth Theory: look it up' stickers I would see on my walk from Chapeltown in Leeds to the University of Leeds, I searched for the term 'plandemic' online. Finding only debunking videos on YouTube, I turned towards Dailymotion – a video streaming site with less regulation than YouTube. Finding what looked like the plandemic video, I clicked play and was immediately struck by the aesthetic, the tone of the interviewer and the stock photography. It had the feeling of a parody video rather than the real video that I was searching for. So, I closed down the video and kept on searching for the official one. After ten minutes, I had circled back to original video that I had started watching earlier. It turned out this *was The Plandemic* production that I had been searching for – so I pressed play again and watched the video, completely fixated, all the way through to the end.

What can this experience tell us about using digital data to derive people's beliefs in misinformation? In short, it tells us that 'a view' (amalgamated into millions of views) is not a great indicator of the *experience* of viewing (Cheney-Lippold, 2017). In fact, it squashes the multitude of different experiences into a single figure. Personally, I watched the video more as a cultural phenomenon than I did for its informative content – maybe in a similar way to the author of *The Telegraph* article. I was not converted to the tenets of the production or any other conspiracy theories.

But another, more technical note, should also be addressed. I watched the video at least twice. While Dailymotion does not have the same 'views' function as YouTube, if I had watched the video during its early stage on YouTube (for at least 30 seconds) I would have added +2 to the total number of views. True, most back-end systems have the capability to distinguish between a 'unique view' (a 'new' viewer) and the total number of views, but this is not what is presented to us on YouTube or what was cited by the article reporting on eight million views for *The Plandemic* video.

This touches on contemporary debates in media and communication studies. C. W. Anderson calls this 'third generation media effects': the adoption of data science tools, often involving data scraping (where vast banks of digital data are collected automatically), to collect large amounts of digital data that can then be analysed with the assistance of artificial intelligence (often referred to as machine learning) (Anderson, 2020a, pp 2–3). This works to create a particular idea of the public, one that can be applied to the idea of who watched *The Plandemic* video.

The person is an individual who recorded their interest in misinformation through a view on a video-sharing site, this view represented their interest *and* belief in misinformation and internalizing this belief would most likely lead to 'mis-behaviour' that would present a considerable challenge to public health efforts. As I hope this chapter has demonstrated, this explanation is too simplistic and empirically dubious.

A strategic emphasis on quantitative realism

Does this mean that we should reject all forms of quantitative knowledge about the public? This is not what Anderson is arguing for in his article and is not what I am arguing for in this chapter. Instead, I argue two things. The first is that survey data should be critically engaged with. There are some great survey designs that acknowledge the constraints of this research method and attempt to mitigate them. But there are also others that are overly simplistic and therefore cannot tell us much about people's beliefs at all.

The second emphasizes the importance of certain types of digital data that does not over-reach. We need to treat this data for what it is, a great example of which can be found in the earlier section about the public's adherence to lockdown rules. The movement data presented by the government in early April involved four different strands of data from The Department for Transport: all motor vehicles, Transport for London Bus, Transport for London Tube and National Rail (the train system that covers the UK). Each involved a slightly different methodology.

For road traffic usage, 275 representative traffic count sites across Great Britain are used to estimate national changes in motor vehicle traffic. For National Rail estimates, the national revenue settlement service provides information on almost all ticket purchases – these are used to organize the data into certain days and certain services. For Transport for London tube and bus routes, the data from the tap-in-tap-out system was used. Each of the four measures compared the current day or week with the equivalent day (or week) in 2019 – effectively saying how much less busy 2020 was compared to 2019 for national motor vehicle usage and train travel and London tube and bus usage (Department for Transport, 2021). This is by no means a perfect way to measure the change in the movement of people, but it operates on a different premise from the digital data about *The Plandemic.*

It is used to assess a very specific and limited phenomenon – whether people are moving about less during a period of national lockdown that is aimed at reducing the movement of people and the spread of the virus. In this way, movement data captures more closely the phenomenon it attempts to measure. It does not attempt to say, 'movement data has reduced so people trust the government' or 'people are moving about less because people don't believe in conspiracy theories'. The criticisms of conflating what people *do* with what people *are* or *believe* is null and void, because the *doing* is the end itself for this data. Such an explanation is also instructive for Google's mobility data. While not a perfect measure of where people go (think of all the people with their location settings turned off), it is used to estimate what people are doing rather than what they are thinking, believing or aspiring to be.[10] It is using digital data in this way that constructs a quantitative realism that is more convincing.

Despite this, it is the data that over-reaches that is given significance within the infodemic. When we consider the way journalists used these numbers, we can observe a *deliberate* over-reach of data to emphasize their own trustworthiness. The narrative goes: 'the infodemic is rife – eight million people watched *The Plandemic* and 7 per cent of people don't think there is evidence for the pandemic – so now more than ever we need journalism that provides credible information to the public.' Here, journalists needed quantitative realism to shore up the boundaries of the infodemic.[11]

The importance of constructing this quantitative realism seems to outweigh the need for critical engagement with the empirical evidence. It ignores the high levels of public adherence to the first lockdown, as shown by the available evidence, and instead emphasizes the threat of the infodemic. In doing so, they miss the broader significance of misinformation: how it is the product and the facilitator of a hyper-sceptical public that are distrusting of the information they receive from politicians, journalists, experts and celebrities.

This lack of critical engagement with the two statistics that represent this threat, means that the news media – intentionally or unintentionally – put forward misinformation about misinformation. That is, they consistently put forward a misleading narrative about the infodemic that is supported by considerably flawed data. Such an uncritical engagement with data becomes more alarming when we consider the excellent technical appraisals of official statistics by journalists. If statistics on the number of infections and deaths receive such excellent interrogations, then data about the infodemic should too.

In facilitating the data bound of the infodemic, the news media's efforts have an unintended consequence. By emphasizing the scale of the infodemic, and its threat to public health, the news media shift responsibility away from authorities towards the public. The conversation stops being about the slow response by the UK government in March 2020 to one that centres on the spread of misinformation online and its uncritical belief by a naive public. And the importance of the former should never be forgotten.

The government was told by SAGE in late February 2020 the necessary measures to the tackle the pandemic: shut down all but essential businesses and state services and tell the public to reduce their social contacts by only leaving the house for necessary activities. But they decided to lock down just under one month later on 23 March 2020. Whatever the reason for this delay, the effects of such a sluggish response is clear – more people were allowed to become infected, more people developed long COVID and more people died from the disease.[12] It is political decisions, not misinformation, that drove the UK death toll so high – if any other narrative seeds itself, it will weigh heavily on the shoulders of journalists that consistently constructed the infodemic rather than challenging it.[13]

6

Data Bounds Are Emotive

Whereas the previous three chapters have focused on the relationship between quantitative realism and data bounds in general terms, Chapter 6 and Chapter 7 circle back to the *Trade-Off* data bound introduced in Chapter 2. This chapter focuses on a data visualization that came to visually represent *Trade-Off*: the graph showing the number of cases, hospitalizations or deaths per day across the entire pandemic. This 'humped' graph – capturing how cases rose and fell across 2020 – provides a way into a discussion about the affective qualities of data visualizations – and by extension, the emotive nature of data bounds themselves. It does so by tracing the story of a particular performance of this graph by a Sky News presenter.

On 11 November 2020, the UK passed the grim landmark of 50,000 deaths within 28 days of a positive test for coronavirus. Later that day, Sky News released a two-and-a-half minute video on YouTube titled 'COVID-19: How did the UK get to 50,000 deaths?'. The broadcast was relatively simple: a journalist, Roland Manthorpe, stands in front of a large screen containing a succession of data visualizations. He begins on the right of the screen, moves to the left part way through and then comes back to the right again – all the while expressively using his hands, posture and voice to provide his interpretation of the changing images behind him. Nothing about the components of this clip is particularly unusual – presenters will often stand next to, or in front of, a data visualization and explain it to the public.

But it was *how* Manthorpe performed that underpinned most of the comments below the video. One comment by Jake Jabz read: 'Why is he so animated, he's talking about deaths in the UK like he's a presenter on Blue Peter' (Sky News, 2020c).

For Jake Jabz there was too much animation for the severity of death, and this resulted in a performance closer to children's television (*Blue Peter*) than a serious news broadcast. And there is something true in this comment – the first time I saw this clip, I was struck by the oddness of Manthorpe's approach to telling this data story. It all felt a bit too energetic, lively and affective for graphs about deaths.

My feelings after watching this video, and the views expressed in the comment section, draw our attention to the emotional conventions in the performance of data visualizations. To unpick *how* visualizations, emotions and performance come together, the chapter takes a 'genre studies' approach to the Sky News video.

Genre studies encourages us to see 'genre' as a relationship between media texts, the producers and the audience (Tolson, 1996, p 92). If we take the broad genre of 'news', we should see news media items (print newspaper articles, digital news pieces, broadcast productions, and so on) as conforming to an agreement between the producers and the audience of news. When we watch the news, we know we are watching the news without anyone having to tell us: it has the structure of news, the aesthetic of news, the feeling of news and the content of news. But the genre of news is not fixed. Both the producers and the audience have a power to change what falls within the genre. The industry can play around with the idea of 'news' by introducing new technologies to news productions, such as visual interactives. But, if this change in production goes too far, the audience can push against the idea that the production is 'news' at all.

But the audience is not just reactive to changes. They can push for news to include topics that are not traditionally associated with the genre, for example, reports about celebrities. But whether the news industry includes this topic, is also largely dependent on the financial viability or public value of such a decision. Therefore, we can see how this relationship between the audience and the producers is set within broader shifts in cultural norms, technological advances and other structural dynamics (such as funding or political pressures).

An analytical use of 'genre', therefore, is useful to identify what we should *expect* from a particular media text – with that *expectation* being a constant negotiation between the producers and the audience (Buozis and Creech, 2018). The comment by Jake Jabz, and other comments in the comment section for the video, pointed to the way this expectation was not met.

The disconnect between audience expectation and the news clip is the starting point for this chapter. From this position, it asks two questions: how did this news clip flout convention? And what does this say about data bounds more broadly? To answer these two questions, the 2½-minute clip is placed within a mini-genre of Sky News clips: where a reporter presents data visualizations about deaths during COVID-19. To outline the conventions of this genre, the elements of this video – the case study for the chapter – are traced back through previous videos from March to November 2020. These other videos are used alongside the case study to outline the data visualization and data performance conventions.[1] Through this, we can see exactly how Roland Manthorpe flouted convention.

Data visualization

In the 2½-minute clip, there were five different data visualizations included. The first involved a bar chart that plotted the number of deaths per day that occurred within 28 days of a positive test, covering early March 2020 to the start of November 2020. Here we can see the distinct 'humps' of the two waves of coronavirus in the UK – one from March to late June and the second from mid-October through to 5 November (the cut-off date for the graph). The second and third graph used the same bar chart template but used the data from the ONS, NISRA and NRS to outline the number of deaths per week broken down by setting (for example, hospital, care home, private home) and age (for example, 75–84 and 85+). The fourth and fifth graphs moved away from the time-series bar chart towards a horizontal bar chart. These two graphs took all of the COVID-19 deaths, as recorded by the ONS, and provided a breakdown by age and ethnicity.

These five graphs – while taking slightly different forms – followed closely the type of charts used in the other Sky News pieces. In particular, the first three graphs were indicative of a convention within this mini-genre that was observed from March to November 2020. Along the *x*-axis was time, beginning at the start of the pandemic (around the start of March) and extending through to the latest data point. The time period was either split by day or by week. Along the *y*-axis was the frequency of the phenomena occurring. On these *y*-axes and *x*-axes, three phenomena were plotted – either cases, hospitalizations or (as was the case for this video) deaths. These three phenomena were either expressed as a bar chart or a line chart. Whether representing cases, hospitalizations or deaths on either a bar chart or a line chart, the aesthetic of these graphs remained similar – the distinct 'humped' graph that saw the waves of the pandemic expressed in peaks and troughs.

Representing the experience of the pandemic

The ability of this type of visualization to become a key part of the mini-genre reflects the way this image established itself as *the* visualization during the pandemic. We can observe the emergence of this graph in political, academic and news media communication in the UK at the beginning of March 2020. These early images were largely illustrative. They provided two hypothetical trajectories of cases: a steeply rising line with a high peak if nothing was done to tackle coronavirus and a more shallow line with a lower peak if public health measures were introduced (Milman, 2020). Even though these early graphs were illustrative – that is, not using actual data – their 'humped' form was reflected in the emergence of modelled projections in early 2020. Most notably, the Imperial College London (ICL) model from 16 March 2020 outlined five possible 'humps'. The steepest and highest peak resulted from the government adopting a 'do nothing' strategy and the shallowest and lowest peak would occur if they introduced all available non-pharmaceutical interventions (case isolation, closing schools, and so on) (ICL, 2020).

The shape of the illustrative and modelled projections became reality as the pandemic progressed. The first wave of the pandemic in the UK (between March and June 2020) saw the emergence of the 'humped' projection for cases, hospitalizations and deaths. It was during this period that the Sky News clips increasingly adopted these graphs to explain how the pandemic was progressing. They mapped the number of daily positive cases on bar charts, compared the number of weekly deaths to the average over the past five years on line graphs, and plotted the rise and fall of hospital admissions using a seven-day rolling average. Whichever phenomenon was represented, the distinct 'humped' nature of these visualizations was present.

But why did this type of graph come to dominate the way that the virus was visualized? On one level, we can see how deaths, cases and hospitalizations are probably the three key indicators of public health during the pandemic. Not only do they highlight the toll the virus is taking on people's lives, they also provide a rationale for public health decisions. As Chapter 2 highlighted, a steep rise in any of these three phenomena means that the government has to take serious action – in the case of the UK, entering in a national lockdown. This firmly placed the data visualization within *Trade-Off* as it positioned health and economics in opposition to each other.

But there is a range of ways that these public health indicators could be represented visually. Most obviously, they could use the same format of graph but plot cumulative data instead – where the distinct 'humps' are converted into a graph that shows rises, plateaus, rises and plateaus. These visualizations *were* created during the pandemic by the Conservative government in the daily press briefings. But they did not have the enduring quality of the humped graphs used by Sky News. So, what makes the humped graphs special? I argue that they capture a collective experience of *Trade-Off*.

The first big hump represented the confusion of the first wave, the unprecedented public health measures that resulted in a national lockdown and all those sunny spring days hoping that restrictions would ease. The consistently low levels that followed embodied the end of restrictions, the dissipation of the threat, the government policies that encouraged people to go out and spend money and the hopes that the worst was behind us.

The rise of the curve in September and October went hand-in-hand with the fear of schools and universities returning, the leaves falling from trees and the anticipation of the inevitable second lockdown in early November. When we approach the visualization in this way, we can see how the image taps into a cultural truth that talks to the affective realm of the pandemic: trauma, anticipation, confusion, anger and sadness.

The ability of this visualization to capture the experience of *Trade-Off* can be seen as one of the reasons why these charts came to dominate the visual representations of the pandemic. For Sky News at least, it seems that these visualizations were constitutive of the mini-genre itself. But this visualization did more than tap into a collective experience; it also operated as journalists' claim to a mathematic truth. To appreciate how these types of visualizations performed this role, we can turn to the work of Anderson (2020b).

Turning death into a graph

Anderson (2020b) urges us to think of data visualizations as forms of 'abstraction': taking all the complexity and open-endedness of the real world and squeezing it into a numbers-based representation. But how can we understand this process? We can think about the concepts of *selection*, *simplification* and *mathematization* (Lynch, 1988). All three help explain how the quantitative realism of the data visualization was established.

Selection draws our attention to what is being *selected* for the visualization. From all of the different phenomena that could be included on this graph, for example, cases or hospitalizations, what is actually selected? Looking at the graphs we can see how this visualization has selected deaths related to COVID-19 over time. In selecting this phenomenon, there follows the process of *simplification*. How do you take all the deaths from COVID-19 over time and simplify it enough, so it fits on a graph?

We can answer this question in a couple of ways. The first is a technical explanation. In Chapter 2, the tricky technical issues of quantifying deaths during a pandemic were documented. Across all the Sky News clips, the visualizations simplified deaths in a number of ways. They used the 28 days after a positive test definition from the UK government, the clinical judgement provided by the ONS in their weekly report and all-cause excess mortality using the data from the ONS. Each dataset involved its own process of simplification (see Chapter 2 for more details of exactly how).

But the conversation about simplification cannot just be a technical one. We need to consider the abstraction that is inherent in converting the phenomenon of death into a bar chart. Yes, the biological transition from a person being alive to being dead can be counted by statisticians (and such a task is an important one). But to count something does not capture its essence. The deeply psychological, familial and cultural phenomenon of dying escapes the bounds of quantification. It is this aspect of death – every part that extends beyond a strict focus of a biological death – that can never be captured in a data visualization of the number of deaths over time. In part, this is why the visualization only refers to the number of biological deaths over time rather than attempting to quantify the meaning of death itself.

This leads us to the third concept: *mathematization*. In stripping away the quantitatively elusive phenomena of death and focusing on its biological aspect, the visualization positions death as something that *can* be counted. In being counted, it positions the visualization as reflecting a sort of mathematical world order. In the same way you can count rocks at the beach or the number of cups in your cupboard, you can count the number of deaths occurring during the pandemic. The visualization leans on the mathematical order of the world to allow this abstraction of death to feel natural, logical and rational (for more on this, see Chapter 3).

Therefore, the processes of selection, simplification and mathematization are reliant on each other. Phenomena that are easily counted – for example, biological deaths – are selected for data visualizations because it allows the simplification, inherent in representing the complex world in a graph, to appear mathematical. In this process, the visualization – and the Sky News reporter who uses them in their presentation – are attached to a sort of mathematical objectivity that can put forward a strong claim to absolute truth. This association between reporting and factuality is considered one of the pillars of the broader journalistic genre (Enli, 2015; Hermida, 2015; Hearns-Branaman, 2016) – achieved here through the way data was represented.

But Sky News did not just rely on the visualization as a form of knowledge about the world to establish their factuality. Each of the visualizations used in the clip referred to an external source. The source was clearly identified at the bottom of each visualization, referring directly to the UK government, the ONS, NISRA or NRS. The inclusion of these sources is an important element of establishing truth in reporting (Kennedy et al, 2016; Lawson, 2021a). Instead of saying this abstract visualization that captures a truth about the world is produced by 'us', they defer to an external authority. These external authorities are organizations that are trusted to produce reliable and accurate statistics. Therefore, the reference to a source says that this abstraction of reality is a legitimate one because it has been produced by an organization that follows the rules of mathematics and statistics (Porter, 1995).

Taken together, the image contains two interlinked messages: (1) while inevitably an abstraction from reality, this captures deaths over time in a single image; (2) this abstraction can be trusted because it came from an authority on statistics that follow internationally agreed mathematical procedures. Both underpin the quantitative realism of the visualization.

Roland Manthorpe's clip, the case study for this chapter, follows the conventions outlined above regarding the use of data visualizations – the graphs he presents are approximately the same as those presented by reporters before him. Therefore, the flouting of the convention did not occur at the level of the visual. It was the performance of these data visualizations that rendered the clip unusual. To appreciate this, we need to outline the performative conventions within the mini-genre.

Performing data

Despite publishing much of his iconic work in the late 1950s and early 1960s, the work of Erving Goffman remains key when approaching 'performance'. He provides a useful way of seeing performance as split between front-stage and back-stage. In our context, the front-stage is the journalist presenting the data visualization on camera and the back-stage is the part the audience cannot see – where the journalist prepares to perform (Goffman, 1959). For the purposes of this chapter, we will focus on the front-stage of this performance. This means looking at how the journalist uses their voice and their body when presenting – much in the same way we would evaluate the performance of the theatre actor.

But this performance should not be seen as hyper-individualist – the journalist does not act exactly how they want to. They are specifically told how to act in the 'back-stage' by news producers and colleagues, but they also learn how to act by observing other people's 'front-stage' performances. In doing so, the journalistic performance is, more often than not, a reproduction of a norm of how to present the news.[2] Or, in the language our analytical framework, the journalistic performance becomes symbolic of the performative norm within the mini-genre. This norm is established through a succession of performances of data that are deemed as unacceptable or acceptable within the agreement between the news producers and the news consumers (Borden and Tew, 2007).

In our numerical case study, we can observe how Roland Manthorpe's performance flouted this convention. The reasons for this (found in the back-stage) are not of concern here. What is of interest is *how* this performance flouted the convention of presenting data visualizations on death by Sky News.[3] To understand this, we will first consider the 'normal' way to perform by examining the news clips that preceded Roland Manthorpe's presentation. To do so, we can think of these performances in terms of voice and action – distinct entities, but ones better understood in tandem than separately.

Performance norms in Sky News

The data visualizations themselves told a story within the humps of cases, hospitalizations and deaths. But the voice and action of the journalist provides a specific narrative that ties together these broad cultural understandings with specific interpretation of the latest data. We can see how this was done in a relatively standardized manner by taking an example.

In a YouTube clip from 9 June 2020, a Sky News reporter opened the piece by facing square to the camera in front of a standard 'VIRUS PANDEMIC' opening slide and explained that the 'latest data' had just been released. The transition from the 'VIRUS PANDEMIC' slide to the first data visualization was reflected in a shift in performance. They would turn to the data visualization behind them to point at different aspects of the graph, while they explained technical details, the scale on each axis and the phenomena being quantified.

> 'So, as you can see this is the average that have died, starting in January and then going through to where we are at the moment.'
>
> *Points at graph and moves finger from left side to right side.*
>
> 'But let's add on the line for 2020.'
>
> *Walks from the left to the right of the screen.*
>
> *[SLIDE: overlays 2020 line for deaths on top of the average of 2015–2019]*
>
> 'There you can see pretty vividly, what happened with COVID-19. The crisis increased the number of people dying across the UK. And I should say this is deaths from all causes.'
>
> *Points at 'hump' in graph where deaths peaked.*
>
> (Sky News, 2020a)

But their performance went beyond merely describing the characteristics of the graphs being presented. They also engaged in interpretation. This involved explaining why different slides were technically distinct from others and what they could tell us about the pandemic itself. We can see how they played out in the extract below, taken from a video clip published on 10 April 2020 titled 'Could the UK be heading for 1,000 coronavirus deaths a day?'

> 'But it's difficult to compare the rate at which they are rising using these lines.'
>
> *Points to lines on the graph.*
>
> *[SLIDE: transition from linear line chart to a logarithmic line chart]*
>
> 'So, let's show you that same data in what we call a logarithmic chart. This essentially allows us to look at the speed of growth …'
>
> *Points to lines on the graph.*
>
> '… so, at first those lines rising quite steeply there and in recent days levelling off a little.'
>
> (Sky News, 2020b)

We can see in this extract how slides transitions, voice and action are all enmeshed within this interpretive performance of the journalist. This interpretation is not just a technical one – discussing the difference between a linear and logarithmic graph – but also in the significance of the data – how this logarithmic chart shows that the rate of increase of these deaths has reduced. This type of performance – the descriptive and interpretative role of journalists – is integral to this mini-genre. But this norm is largely reflective of the wider genre of presenting data visualizations. So, what made the presentation of data visualizations about death distinctive?

Sombre performance of death

The key lies in affect. In one of the first video clips, Ed Conway – a Sky News presenter – opened his performance by explaining:

> 'If you know me, you know I spend a lot of time in front of this wall. Bouncing around. Looking at the economic data. Looking at the graphs. Getting quite excited by them. But I'm afraid the graphs we have here for COVID-19 are deeply depressing. Some of the most depressing I've ever seen. And let me explain why.' (Sky News, 2020d)

This statement reveals the difference between presenting deaths during the pandemic and other types of data. The excitement and the animation that characterize other mini-genres are replaced by a performance that is more befitting of the 'deeply depressing' nature of the data. We can observe the way this played out in the subsequent news clips. The journalist was still engaging in the descriptive and interpretative performances described above but they made use of their movement and their voice quite differently.

This did not mean that the journalists were consistently monotone, completely inexpressive and entirely stoic. But when they did use their voice and action in their performance, it was notably muted and sombre. There were certain ways that journalists *never* spoke during these data visualizations – they did not laugh, they did not adopt ironic commentary and they were not playful in the data they described. Instead, they spoke about the visualizations in grave tones that expressed the severity of the numbers they were describing. When more expressive tones, pitches and intonations were used, they were largely tied with the changes or significance of the data visualizations they were presenting.

A similar pattern emerged in the *action* of reporters. There were definite no-go actions, such as overly expressive facial expressions or wild gesticulation. In general, the reporter would stand in the same spot and interchange between talking directly to camera and pointing at the relevant parts of the visualization on the screen. When the reporter did move from their initial position, they would move in slow, purposeful strides that would take them either to the other side of the screen or closer to the data visualization (so they could point to specific details). Similarly, when the journalist went beyond pointing to the graphs, they would either engage in hand-wringing – conveying the severity of the data they were presenting – or open up both their palms to mirror their explanation of the graphs.

If we return to the example cited above regarding logarithmic graphs, we can integrate these aspects into the description of the performance with the ⋆ referring to moments when the reporter ⋆emphasized⋆ a particular set of words:

> *Open palms moving as they talk.*
>
> 'But it's difficult to compare the ⋆rate⋆ at which they are rising …'
>
> *Points to lines on the graph.*
>
> '… using these lines.'
>
> *[SLIDE: transition from linear line chart to a logarithmic line chart]*
>
> *Open palms moving as they talk.*
>
> 'So, let's show you that ⋆same⋆ data in what we call a logarithmic chart. This essentially allows us to look at the ⋆speed⋆ of growth …'
>
> *Turns 90 degrees towards the visualization, small step towards the image and points to lines on the graph.*
>
> '…so, at first those lines rising quite steeply there and in recent days levelling off a little.'
>
> (Sky News, 2020b)

It was in this context that Roland Manthorpe presented the data on deaths in November. His performance, however, was markedly different from the sombre performance of the other news presenters.

Flouting the convention

Roland Manthorpe performed his descriptive and interpretive role in much the same way as the other journalists in this sub-genre. He outlined the technical aspects of different visualizations and provided a clear answer to the question that formed the title to the clip: 'How did the UK get to 50,000 deaths?' But it was the way he used his voice and his body that pushed him outside the convention of presenting data visualizations about deaths during the pandemic.

Whereas many other presenters stayed still or moved with slow, purposeful action, Manthorpe was far more animated. In the first 30 seconds, he had moved from the left of the screen to the right, and over the next two minutes moved from the right-hand side to the middle, and back again, multiple times. But it was not just how much he moved; it was also the *way* he moved. He shuffled towards the graph and away from it, turned 180 degrees with speed and often crouched to draw attention to specific areas of the graph. This animation was occasionally reflected in his facial expressions too: the shift from furrowed eyebrows to raised eyebrows when explaining confusing parts of the data. It was this type of animation to his actions that set him apart from other reporters in this mini-genre. This was matched with a certain way of using his voice too. At the beginning and the end of the news item, he uses two rhetorical questions:

> 'How did we get to 50,00 deaths? The simple answer: with shocking speed, in one month, in April.
>
> We know when and we roughly know who, but of course the question we are all asking is why? Looking back at this data, it is hard not to see the lockdown in March as the crucial moment.'
>
> (Sky News, 2020c)

The use of these questions added a suspense to his presentation that was largely absent from other performances. Twinning with these rhetorical questions was the use of adjectives (for example, 'stunning'), pauses between words and sentences, the deliberate speeding up of speech, stresses on certain words over others and a sort of breathlessness that characterized his entire performance. Taken together, the voice of action provided a performance that pushed at the edges of convention. We can see this in the transcript below where [pauses] have been put in square brackets and moments where he sped up his speech identified through carets (^).

> 'Pre-existing conditions were also incredibly important.'
>
> *Closed fist with the thumb on top (the classic politician pointing gesture).*
>
> *[SLIDE: transition to visualization of pre-existing health conditions split by age]*
>
> '*Most* people who died had a pre-existing condition [pause] that [pause] is these light blue boxes here... .'
>
> *Points to light blue boxes.*
>
> 'Although [pause] interestingly [pause] that was less common among younger people.'

Moves to centre of visualization, crouches and points to lower left part of visualization.

'We don't [pause] ^fully understand that^ ...'

Raises eyebrows as he exits a crouching position.

'... as with so many other things.'

(Sky News, 2020c)

For many in the comments section of the clip, this level of expression provided a news performance that was closer to a children's television programme than a Sky News presentation about data visualizations on death during the pandemic. While I do not agree with the specific (and somewhat harsh) criticism, I would agree with the essence of these comments: this was a performance that flouted the mini-genre that had been established from March 2020. This was not done through the data visualization or the content of his description and interpretation. Instead, it was through the way he used his body and his voice. Something that appears so minor had a considerable impact on the validity of the performance itself.

Feeling data

The affective quality of the quantitative is often neglected in scholarly work, which focuses on how numbers intersect with rationality, computation and understanding. However, Kennedy and Hill (2018) emphasize the need to attend to '*the feeling of numbers*, or *how numbers feel*' in the age of datafication (Kennedy and Hill, 2018, p 844). This case study provides one way into the emotive nature of the quantitative.

By taking a genre studies approach, the chapter highlighted how the audience expected the 'humped' graph to be presented by journalists in a particular emotional tone. Roland Manthorpe gave an animated performance that involved too much emotion, expression and feeling compared to other performances within this mini-genre. Therefore, he flouted the need for seriousness that the data visualization necessitated.

But explicit displays of emotion in response to horrific topics are not always treated as absurd. Borden and Tew (2007) highlight how overt displays of emotions by reporters witnessing Hurricane Katrina, reporting on 9/11 or the Hindenberg crash were justified by the 'scope of the disaster, and the government's failure to respond to it adequately'. In these contexts, it is more acceptable to display emotion and gesticulate than to remain sombre and detached. Considering that the pandemic is comparable to these crises and disasters, it cannot just be the topic that restricts emotion. There is something in the nature of the data itself.

As Kennedy and Hill emphasize, data can elicit emotional responses. But it also involves securing quantitative realism through selection, simplification and mathematization, as described in the earlier section of this chapter. This requires a considerable degree of abstraction, where the colossal phenomenon of death is turned into impersonal quantitative information that can be displayed on a graph. This visualization can then be communicated to the public outside the immediate context of the event. For journalists, this involves being 'in the studio' rather than 'at the scene'. Therefore, this mini-genre was a far cry from the iconic reporting of disasters. Journalists take on the role of the sombre statistician or scientist, presenting data as detached and rational individuals looking to inform the audience, not excite it. In this context, overly emotive performances about these types of graphs – even when they are about something as affective as death – are not acceptable.

So, we can see how this data visualization gained cultural significance as an affective piece of imagery. But what does this story say about data bounds more broadly? The concept of data bounds is deliberately broad: where the quantitative becomes a meaningful way to engage with, discuss, respond to, think about (and so on) specific phenomena. In respect to this chapter, we can see data bounds as a cultural moment where the quantitative becomes a meaningful way to *feel* the pandemic.

The data visualization achieves this by effectively representing the *uncertainty* of living within *Trade-Off* by capturing the waves of emotion associated with the peaks and troughs of cases, hospitalizations and deaths. And it also provides a clear indication of the widespread *effect* of the pandemic on people, from experiencing long COVID to dying from the virus. This allowed the graph to operate as a gateway for people to emotionally engage with the history, present and potential future of the pandemic. The ramifications of this are explored in more detail in the final chapter. But before we engage with this, the following chapter provides the final perspective on data bounds.

7

Data Boundaries Are Drawn Within Historical Norms

Understanding data bounds is not just about data. This chapter marks a break from the previous five by identifying how broader historical norms can help shore up certain data bounds and marginalize others. It does so through a single case study: the way a projection of 90,000 cases per day failed to circulate in the media.

In May 2020, a group of experts in the UK set up the Independent Scientific Advisory Group for Emergencies (Independent SAGE). This operated in opposition to the official Scientific Advisory Group for Emergencies (SAGE) that advised the UK government on the handling of the pandemic. According to Independent SAGE, they are 'a group of scientists working together to provide independent scientific advice to the Conservative government and public on how to minimize deaths and support Britain's recovery from the COVID-19 crisis' (Independent SAGE, 2021b). Much of their work on communication centres on influencing the news media's coverage and more direct forms of communication (for example, social media) to pressure the government to change their approach to coronavirus.

As part of this strategy, they held weekly press conferences that were live streamed on YouTube. In one of these press conferences on 2 July 2021, Professor Christina Pagel – a member of Independent SAGE – made the following projection: 'In mid-July we will have a seven-day average of 90,000 [cases per day]' (Indie SAGE, 2021). This average of daily cases would far surpass the peaks of any of the previous waves.[1] The alarming figure was used by Pagel to underpin her argument that the government should change tack. Instead of opening society on 19 July 2021 – the proposed date – they should wait until the vaccination programme had been rolled out fully. This would mean people gained immunity through vaccination rather than natural infection.

The importance of the 90,000 figure, and the underlying argument, seemed to land with the two main audiences of the broadcast. Members of the public watching the briefing provided some direct reaction to the figure in the live chat:

> FSM is the dog: '80k … ugh'
> PJ Neil: 'Our own ski jump! :S'
> Steven Corder: 'Yes! Yes! My own modelling indicates that we'll get to about 96,000! I'm well chuffed about that!' (Indie SAGE, 2021)

It also held some sway for the more specialist audience. Nigel Nelson of *The Sunday People* – a British tabloid newspaper – opened his line of questioning by referring to 'those rather frightening figures from Christina' and leading into the implications of this projection on government policy (Indie SAGE, 2021). Beyond the YouTube broadcast, however, the 90,000 figure really struggled to circulate within a wider public discourse.

The projection was published in *The Sunday People*[2] on 4 July but was not a prominent story (Yumpu, 2021). On the same day, a relatively small political magazine called *PMP* directly quoted Pagel's projections in their provocatively titled article 'The government has given up' (PMP News, 2021). Three days later, on 7 July, it also featured in an opinion piece in *The Guardian* by Deepti Gurdasani, a clinical epidemiologist and senior lecturer in machine learning at Queen Mary University of London. She explains: 'Infections are now surging among young people. Given the current doubling of cases every nine days, Independent SAGE predicts that we are on track to see 90,000 daily cases by 19 July, with one million new infections occurring by then' (Gurdasani, 2021).

But this level of attention is sparse, especially when we compare it to the way certain figures in this book gained widespread public attention. On the face of it, the 90,000 figure had all of the necessary components: a certain 'hugeness' akin to 'one billion items of PPE' that normally propels a number through public discourse; this hugeness could be related to immediate risks of rising cases, namely placing lots of pressure on the NHS and leading to large numbers of deaths; it was communicated by a well-respected Professor of Operational Research at University College London (UCL); and this expert, and her claim, formed part of the larger body of Independent SAGE – representatives of which made regular appearances in mainstream media from its inception in May 2020 (Cheng, 2020; Davis, 2020; Woodcock, 2020). Despite this, there was a distinct inertia of the 90,000 projection. Appreciating why this occurred is exactly what makes this number so fascinating.

This chapter argues that we need to place the figure within *Trade-Off*. Then we can understand how the trade-off between health metrics and economic indicators was not concerned with inequality. The 90,000 cases per day would have a disproportionate effect on those living in the most deprived areas of the country – but this did not enter into the equation when deciding on 'opening up' the country on 19 July 2021. We can explain such a situation by looking to the historical normalization of health inequalities over the last 150 years (and beyond). Set within this history, it was logical that a conversation about trading between health and the economy would not consider health inequalities. Before we delve into this argument, however, we need to consider the most immediate reason people would give for the projection not taking off: the inaccuracy of the projection itself.

Red-herring of inaccurate projections

The projection provided by Pagel was relatively simple. At its heart was the seven-day moving average. This figure accounts for the previous seven days' worth of data and presents a mean average. This involves adding all of the data for the previous seven days and dividing this number by seven. This calculation is particularly important when we consider the large fluctuations of cases across a week due to reporting delays over the weekend. Due to this, the figures for Sunday and Monday are generally lower than those on Tuesday and Wednesday. The seven-day average 'flattens out' these fluctuations, providing a figure that can be compared to the previous seven-day average. If we look at the data that Christina Pagel would have used on 2 July, we can see number of cases for the previous seven days:

Friday (25 June): 17,764
Saturday (26 June): 15,961
Sunday (27 June): 18, 123
Monday (28 June): 27,746
Tuesday (29 June): 27,588
Wednesday (30 June): 28,319
Thursday (1 July): 28,104[3]

Adding all of these numbers together, we get a total of 163,605 cases occurring in the previous seven days. If we divide this by seven (the number of days), we arrive at the seven-day rolling average of 23,372 cases. If you perform this calculation for all the previous daily counts of cases, you can provide a moving seven-day average for the entirety of the pandemic. This data allows you to see how the seven-day average changed over time.

There are lots of ways of representing changes over time, but Pagel was particularly interested in the 'doubling time' of cases: how long it would take for cases to double in number. If we took the 23,372 seven-day rolling average we calculated, we would look back in time for when cases were closest to 11,686 (half this number). This date was 22 June, nine days before 1 July. Therefore, the doubling time was nine days.

Pagel then applied this doubling time to the future. She argued that in nine days' time, on 10 July, there would be around 45,000 cases per day. In another nine days, on 19 July, there would be around 90,000 cases. While this approach seems logical, her projection was considerably different from the actual seven-day rolling averages for 10 and 19 July.

10 July

- Prediction: 45,000 cases
- Actual:
  - 33,725 cases by reporting date
  - 35,918 cases by specimen date

19 July

- Prediction: 90,000 cases
- Actual:
  - 39,818 cases by specimen date
  - 46,460 cases by reporting date

(UK government, 2021)

Given the substantial discrepancies between the projections and the actual data, it is tempting to say 'this projection did not receive much attention because it was a poor projection'. This is certainly the approach taken by a *Spectator* article from 27 July 2021 (Steerpike, 2021).[4] To a certain extent, this logic makes sense. Pagel's assumption that cases would keep doubling every nine days soon stopped reflecting the actual data. On 4 July, just two days after her presentation, the data showed a rough doubling time of 11 days – meaning that cases were doubling at a slower rate. This increased to 13 days on 8 July.[5]

But such an argument is unconvincing. We can see how projections do not need to be particularly sophisticated or accurate to circulate. On 17 March 2020, Patrick Vallance – a prominent government scientist – told a committee of MPs that a death toll under 20,000 would be a 'good outcome'. This received a large amount of attention across news media sites (Donnelly, 2020; Heffer, 2020; John, 2020).

This figure was taken from the ICL modelling that was published the day before, on 16 March 2020. In this report, they argued that 20,000 people would die if the government adopted more stringent measures (including quarantine, social distancing and closures of schools, universities, cafes, restaurants, pubs and bars). This was compared to the 260,000 people who would die if the government took no action. This projection proved to be a gross under-estimate. The UK's official toll (missing a considerable number of people who died outside of hospital) surpassed 20,000 on 25 April 2020.[6] Despite this, the figure circulated widely in the middle of March 2020 exactly because it could not yet be verified as accurate or not.

This points to the way the media will be drawn to publishing a shocking figure – one that circulates for a short period of time.[7] So, it is more instructive to consider projections without technical hindsight. It is not about whether they prove to be correct or not. Rather, it is more about how they gain a significant amount of attention in a very short time because they are considered credible, align with certain political beliefs and operate to shock and sensationalize. So, we need to place the 90,000 figure within the context of its immediate communication to understand why it struggled to take off.

Vaccines, cases and risk

If the 90,000 projection was made during the second wave of the virus during autumn and winter (2020/21), it would be considered apocalyptic. The peak of the second wave saw a seven-day average of 61,230 cases by specimen date – these case numbers were substantial enough to force England into the longest national lockdown of the entire pandemic. But this projection was not made during the second wave, it was made in the middle of the third wave in the summer of 2021.

From late 2020, the NHS had rolled out a highly successful vaccine programme. By the time of Pagel's presentation on 2 July 2021, 85.3 per cent of people aged 18 or over had received one dose of the vaccine and 63.2 per cent had received two doses (UK government, 2021).[8] The effectiveness of the vaccines meant that the relationship between cases, hospitalizations and deaths had changed substantially.

Let us work with a hypothetical set of 1,000 people infected with coronavirus. In the second wave of the pandemic, before the rollout of the vaccine, you would expect around 100 of these infected people to end up in hospital. Of these 100 people in hospital, around ten of them would die. By June 2021, these numbers had substantially reduced. For every 1,000 infected people, you would expect 35 to be hospitalized and one of them to die (Sparrow, 2021).[9]

It is hard to overstate the significance of these numbers. The successful vaccine rollout had made it less likely that someone who contracted coronavirus would need emergency care, and much, much less likely that that person would die.[10] The two most important public health metrics to describe the risk of coronavirus cases – hospitalizations and deaths – no longer held the sway they did in the first and second waves. This meant that 90,000 cases a day would result in a manageable level of hospitalizations and would translate into an acceptably low number of deaths. It was this change in the epidemiological formula that underpinned the English government's plan to lift more restrictions on 19 July – scrapping the one metre rule and the wearing of face masks, and allowing the opening up of large venues, such as nightclubs, music venues and theatres (Morton, 2021).

This meant that a once shocking number – 90,000 cases – no longer held the weight that it once did. Such an explanation goes a long way to explaining why the number did not gain a foothold in public discourse. But it is important to understand that the risk of cases is not just confined to hospitalizations and deaths. There are a number of other risks that are associated with cases that mean the 90,000 figure should have offered some concern to the wider public.

During her presentation, Christina Pagel outlined five of these risks: (1) while the link was weakened, large volumes of cases would still result in noticeable hospitalizations and deaths; (2) rising hospitalizations would have a considerable impact on already exhausted and overworked medical staff; (3) each new case of the virus presents a risk of long COVID; (4) new cases will mean continued self-isolation in education and work, leading to loss of education and income; (5) the exponential rise in cases provides a hotbed for variants to emerge (Indie SAGE, 2021).[11]

In light of this, Pagel argued that the government should not go ahead with their plans to relax rules from 19 July. Instead, they should delay this date until all adults in England had been offered, and taken up, their second vaccination. Pagel argued that 'we really need to get to 85 per cent of the population double vaccinated for Delta'. This stood in stark opposition to the government's current strategy of allowing people who are not fully vaccinated to acquire immunity through infection – a process that exposes this group to all of the associated risks of contracting coronavirus (Indie SAGE, 2021).

So, the risk associated with the 90,000 figure can be split into broadly two camps. Approach one argued for restrictions to be eased because the link between cases and hospitalizations and deaths was substantially weakened. Approach two argued for the completion of the vaccination programme before further easing of restrictions due to the risks of hospitalizations and deaths, other hospital care being stretched, staff pressures, long COVID, lost work and school days due to self-isolation and the potential for new variants.

The risks outlined by approach two were not unknown to the UK government and the mainstream news media – who were following approach one – but they were not considered substantial enough to change tack. Why was the government not forced into delaying so-called 'freedom day'? The most convincing answer comes when we consider the unequal effect the pandemic has on society.

Normalizing health inequalities

All five of the risks outlined by Pagel are not felt equally. Pagel was keen to stress that people living in the most deprived areas were more likely to be hospitalized and die from rising cases, more likely to get long COVID, more likely to live in areas where their nearest hospital was experiencing extraordinary pressure, more likely to be missing school and work due to self-isolation and more likely to live in an area where variants would emerge. This claim is backed by a substantial body of empirical research.[12]

This unequal impact of coronavirus is reflective of the pre-pandemic world. ONS data from 2018 highlights the difference in life expectancies at birth between the least and the most deprived areas.[13] Males in the least deprived areas would live for around 9.3 years longer than males in the most deprived areas, while the difference was 7.4 years for females (Finch, 2018).[14] But we should not just see this issue as one about deprivation; it is also a story about ethnicity.

In a study published in 2020, it was shown that Pakistani, Bangladeshi, black African and black Caribbean people were all more likely than the white British population to live in the most deprived decile. In fact, 31.1 per cent of the Pakistani community lived in this decile (UK government, 2020a). This is not to say that this decile did not include white people: the local authority that received the lowest score for deprivation (making it the most deprived) was Blackpool (MoHCLG, 2019) – an area that has a 97 per cent white population, compared to the England average of 85 per cent (CVSBWF, 2013). Instead, it emphasizes that certain ethnic groups had a bigger percentage of people in these poorest areas than the white population. Here we can see the quantitative materialization of Stuart Hall et al's (1978, p 394) famous phrase: 'race is the modality in which class is lived.' That is, ethnicity shapes particular experiences of deprivation but deprivation ties together different ethnicities.[15]

Such disparities in health outcomes have become normal in modern Britain. To appreciate why, we need to trace a longer history of inequality. The work of Engels in the 19th century is often referred to as the starting point for this conversation. In *The Conditions of the Working Class in England*, Engels argues that disease was a result of social organization and, therefore, was socially produced rather than innate to human beings (Scambler, 2012, p 131). Despite his convincing argument, these health inequalities were rarely systematically addressed. Writing in *Data in Society*, an anonymous writer argues that 'inequalities in communicable disease risk in the 19th century were replaced with inequalities in non-communicable disease in the 20th century' (Anon, 2019, p 254).

The post-Second World War period did see a notable improvement in the health of poorer communities, with the introduction of welfare security. Most notably, the creation of the NHS that provided healthcare free at the point of access across the UK. Much of this hard work, however, has been undone in the UK over the past 40–45 years. Whereas some countries have seen health outcomes narrow during this period, others (including the UK) have seen it widen. In the mid-2000s, Coburn (2004) argued that this difference can be linked to the adoption of 'neoliberal' policies. These policies – broadly focused on the reorganization of the economy in favour of less government regulation and more rights, power and money to powerful individuals and businesses set within a 'free market' – were adopted most fervently by the UK and the USA in the late 1970s. In the subsequent years, countries have both seen inequality widen and, as a result, seen health inequalities widen too.[16]

Coburn (2004, p 47) explains that prior to the neoliberal era, income inequality in 'the United Kingdom had been relatively low and declining since the Second World War'. With the establishment of neoliberalism from the late 1970s in the UK, income inequality increased substantially and continued to do so through the 1990s. Not only had it increased within the country, but levels of income inequality were also much higher than other comparable countries that took a more social democratic approach to the economy. This income inequality then manifested in health inequalities, as evidenced in the widening gap between the life expectancy of those at the higher and lower ends of socioeconomic status (2004, p 49). The message is loud and clear: economic policies pursued by governments lead to socioeconomic inequality that, in turn, results in health inequalities.

A wealth of subsequent work has consistently documented and emphasized this relationship between health inequalities and neoliberalism, trade liberalization and the growth of transnational companies; see the work of Navarro and Shi (2001), Navarro (2007), Graham (2007), Chernomas and Hudson (2009, 2010), Raphael (2015) and Doran and Cookson (2019). These disparities were then further compounded by the Financial Crash (2007/08) and the Conservative government's adoption of austerity from 2010 onwards in the UK. This political project implemented cuts in central and local government spending and an increase in taxation, sitting in opposition to governments who chose to invest government money to stimulate economic growth. After reviewing the available evidence, Stuckler et al (2017) explain that austerity had the following effect:

> It impacted most on those already vulnerable, such as those with precarious employment or housing, or with existing health problems. It was associated with worsening mental health and, as a consequence, increasing suicides. Yet, this was not inevitable. Those fortunate to live in countries with strong social protection systems, such as Iceland and Germany, escaped the worst of the crisis.

So, we can see how health inequalities in modern Britain have largely been facilitated by government policy (since Engels' writing over 170 years ago but stretching much further back than this too). They remained as the country shifted from communicable to non-communicable diseases in the 19th century to 20th century, they were further entrenched by neoliberalism and cemented over the past decade as austerity politics was implemented.

Given the structural nature of health inequalities – in how they reflect broader inequalities and how they are entrenched through structural changes enacted by government – then structural policies are the most viable solutions.

These so-called 'upstream' health policies involve making broader changes to society – to address inequalities in power, social status, social connections, income and wealth – which then result in reducing health inequalities. A considerable amount of research highlights how this is the most viable way of tackling health inequalities (Coburn, 2004, 2009; Scott-Samuel and Smith, 2015; McCartney et al, 2019).

But these policies have been consistently rejected by the UK government in favour of 'downstream' policies: attempting to encourage better individual choices about health. These include dissuading people from smoking, educating people to make healthy choices of food and drink and encouraging individuals to use contraception (Scott-Samuel and Smith, 2015). In focusing policy on individuals, rather than structural inequality (or a combination of the two), it has become more 'common sense' to see health inequalities as rooted in a person's choices about their own health (despite all the empirical work that counters such a claim).

This individualistic discourse was highlighted by Rich, Miah and Lewis (2019) in their analysis of government documents about digital health and health inequalities. They found that health inequalities were discussed within the framework of empowering individuals to make better choices. This emphasis on the individual can also be recognized in the research of Mackenzie, Skivington and Fergie (2020) – through 47 interviews with healthcare professionals, they identify a strong emphasis on lifestyle choices as the causative factor for patients' poor health. This need to position health inequalities as individualistic is reflective of the broader notion within political and media systems that poverty is a personal choice – as documented by Lugo-Ocando (2015) in *Blaming the Victim*.

So, we are left with the normalization of health inequalities themselves and the normalization of their causes being individual rather than structural.[17] It is normal for us to see homeless people lying in doorways and normal to see overly grand houses occupied by multi-millionaires. These conditions are considered largely dependent on these people's behaviour and attitude: homeless people are lying in doorways because they are irresponsible or lazy and rich people own grand houses because they worked hard and act as responsible citizens.

The *Trade-Off* data bound established in early 2020 was shaped by this norm. Inequality was not factored into the calculations of different decisions. Within this prevailing reality, there was no specific limit on how badly the virus could affect the poorest in society. Instead, it was accepted that cases, hospitalizations and deaths would be higher in certain poorer areas.

So, we can return to the two approaches outlined above. The first approach provided the 'normal' understanding of health inequalities. It argued that society needed to be opened up because the risk of COVID-19 for the wealthier population (more often white British than Pakistani, Bangladeshi, black African or black Caribbean) had reached a low enough level and the still considerable risk posed to the more deprived populations (less often white British than Pakistani, Bangladeshi, black African or black Caribbean) now fell into an acceptable 'normal' level.

The second approach, on the other hand, argued that this disparity in risks meant that the vaccine programme should be rolled out fully before restrictions were eased. The second approach, outlined in the YouTube broadcast on 2 July by Christina Pagel, was later formalized in an open letter (signed by Independent SAGE members, among other experts) to *The Lancet* – an internationally renowned science journal. Importantly, this open letter does not just argue for the offering of vaccines to all of the adult and adolescent population – it also pushed for the government to make sure uptake of the vaccine was high across all of the population. Given that those in poorer areas were less likely to have the jab, there was a renewed emphasis in this letter on addressing some of this inequality (Gurdasani et al, 2021).

Such a reframing allows us to understand why and how the shocking projection of 90,000 cases a day failed to gain any traction in public discourse. The implicit narrative went something like this. Yes, cases would sky-rocket but hospitals in general would 'cope' and it would not result in a particularly high level of infections, hospitalizations and deaths in richer parts of the country. So, the cases are not threatening enough for us to delay opening up society. The trade-off is worth it.

Set within the context of the past 150 years, such a conclusion makes sense. But the horrors of what such a logic means should serve to shock people: it means that the life of a poorer person is less valuable than that of a richer person. The power of historical normalization is to facilitate the construction of data bounds where that assumption is 'common sense'.

The failed campaign of the 'outsiders'

This chapter has emphasized the need to approach projections without technical hindsight. Yes, the 90,000 cases projection was too simple and proved to be very inaccurate. But other projections that were equally as poor received much more attention. Therefore, there was a push to set these numbers – describing a possible future – within the context of the risk they present. In the case of the 90,000 cases, this highlights how two different approaches can present different interpretations of these risks.

One emphasizes the changed relationship between cases and hospitalizations and deaths as rationale for reducing restrictions on society. The other sees the unequal effects of cases on poorer communities as the reason for delaying the 'opening up' of society until the vaccine programme had been fully rolled out. Both are set within *Trade-Off* but one acknowledges inequality and the other does not.[18]

In providing this context, the reader can then understand why a story about unequal health risks did not capture the attention of the wider political and media system. The substantial difference in the health risks to poorer people compared to richer people has become entrenched in modern Britain through a long history that stretches back to the Victorian period (and beyond). It was normal for poorer people to die earlier before the pandemic and it was normal for the same group of people to be hospitalized more, die more, get long COVID more, miss more days of education and work, and have healthcare facilities under more pressure. So, with the dominant data bound of *Trade-Off* it would seem absurd to not open up society when the effect of the virus was only being felt by the poorer parts of society and not the richer parts.

It is within this space of absurdity that Independent SAGE has spent most of its short existence. Much of the work of Independent SAGE, the group that Christina Pagel is part of, operates on the edges of this normalized view of the pandemic. While the organization itself was seen as reputable, as evidenced by a high number of media appearances, its political and public health view of the pandemic sat on the fringes.

For Cairney (2021), this meant Independent SAGE operated an 'outsider strategy to encourage the critical attention of an external audience' where they rejected the 'rules of the game' played by insiders like its counterpart SAGE. In playing the role of the 'outsider', Indie SAGE spoke 'truth to power' by generating interest from external audiences – the news media, the general public, opposition parties – to pressure the government to change course (Cairney, 2021, p 6).

In some respects, they have been successful. They had a large part to play in the government's decision to reveal the participants of SAGE and the official documents informing, and emerging from, the meetings (Inge, 2020 cited in Cairney 2021). But on more significant policy issues, Independent SAGE has struggled to make an impact. Most notably, they have consistently advocated for suppressive measures to eliminate, rather than manage, COVID-19. But these largely fell on deaf ears (Cairney, 2021, p 10).

Their most recent statement before the 90,000 cases per day projection provides a pertinent example of this – they emphasized the need for supportive measures for those with COVID-19, such as: improved accessibility to vaccination; full pay to get tested and full pay if needing to self-isolate; financial support for those needing to quarantine after international travel;

and a ventilation support fund for businesses. None of these were taken up by the government (Independent SAGE, 2021a). It seems that Independent SAGE are well used to occupying this odd position: welcomed into the mainstream as sources by journalists but rarely having their argument properly heard by those in positions of power. The projection of 90,000 cases stands in for this broader struggle to forge an alternative data bound: a *Trade-Off* that includes inequality or, more broadly, something closer to *Protect Both*.

8

Critically Engaging with Data Bounds

These six short stories point to the increased importance of the quantitative during the pandemic. While much of the pre-pandemic world was dominated by digital data, often describing individual behaviours online, the pandemic and post-pandemic world has forced numbers about society onto the mainstage. Before the pandemic, people would be familiar with GDP or unemployment rates. But the scale, scope and familiarity with this type of data marks a distinct juncture.

This makes understanding the cause, nature and effect of the quantitative a pressing task in the 'post-pandemic' world. To do so, this book argues that we need to engage with *data bounds*: how data becomes a meaningful way to experience, think about, discuss, react to, engage with and change certain phenomena. These data bounds are complex, made up of an ensemble of technical processes (collecting data, cleaning it, analysing it), contexts (political, economic, cultural, and so on) and media and communication.

But this complexity should not inhibit understanding. Each chapter in this book offers a distinct perspective of data bounds: they are reinforced by policy (Chapter 2), quantitative realism underpins them (Chapter 3), quantitative realism is mathematical and abstract (Chapter 4), desire for them underpins quantitative realism (Chapter 5), they are emotive (Chapter 6) and their boundaries are drawn within historical norms (Chapter 7).

These perspectives allow for data bounds to be broken down into six characteristics. In doing so, it can be used to think through non-pandemic phenomena that are highly quantified. These include – but are not limited to – health and fitness, inflation and cost of living, crime and justice and finance. It is hard to discuss, engage, experience or think about these contexts and for data *not* to be meaningful. To speak of the cost of living is to speak of wages, prices and profits. It is not to say that the qualitative plays no role, but to argue that the quantitative dominates.

Just as there are highly quantified phenomena, there are also contexts that have not felt the data creep so acutely. We can think of poetry and literature as existing outside data bounds. Yes, we can look at the rating of our favourite books on Goodreads, decide on our next poetry book based on the copies sold or talk about the number of pages we have read. But this is not a particularly meaningful way to engage with poetry and literature. Instead, we talk of characters, plot lines, cliff hangers, tension, pace and lore in (almost exclusively) non-numerical terms. In the future, we may use biometric data to track just how much our pulse increased at that cliff hanger or data from brain scans to track personal involvement in character development. But for now, numbers are still ancillary to the context of reading books and poems.

For highly quantified phenomena, however, data bounds are a useful critical tool to work with.[1] The final chapter is dedicated to outlining a four-part toolkit to help scholars put data bounds into practice:

1. Pay attention to media and communication.
2. Interrogate and appreciate quantitative realism.
3. Examine how data bounds can maintain or challenge power.
4. Determine why some data bounds dominate over others.

Pay attention to media and communication

There is a wealth of overarching theories of the quantitative in the social sciences and humanities: data infrastructure (Kitchin and Lauriault, 2018), data regimes (Dalton and Thatcher, 2014), data assemblages (Kitchin and Lauriault, 2014), data space/code (Kitchin and Dodge, 2011) and data engines (Kitchin, 2021b). Many of these concepts have come on the back of the data deluge of the 21st century and have – directly or indirectly – forged Critical Data Studies (CDS) as a discipline. These terms, therefore, have crystallized within the dominant fields within CDS: geography, sociology, science and technology studies, and history and philosophy of science. While this work has championed the meeting of the technical and context, it has rarely engaged substantially with media and communication.[2]

This is what sets data bounds apart from the other data-related concepts listed above. It emphasizes the central role that media and communication has in rendering data a meaningful way to engage with certain phenomena. Each chapter is testament to that fact. They can be read as odes to the power of media ecosystems. But we also need something of a schema that provides an overarching understanding of these media ecosystems.

The media ecosystem

We can begin by appreciating the different forms of communication available in the media ecosystem. Each chapter provides a different angle:

- **Discourse:** broad structures of meaning formed of data (Chapter 2).[3]
- **Linguistics and science communication:** language of mathematics used in scientific documentation and publications (Chapter 3).
- **Rhetoric:** strategic use of numbers to convince and persuade (Chapter 4).
- **Representation**: using numbers to create identities (Chapter 5).
- **Semiology and performance:** juxtaposition of data visualizations and the performance of them (Chapter 6).
- **Risk communication:** communicating the risk of an event, process or policy decision for individuals and society (Chapter 7).

This communicative starting point for numbers is important. As each type of communication has its specific characteristics, it structures what meaning is given to the quantitative. Rhetoric is closely tied to the speeches of politicians, where language is used to convince and obscure. That means rhetoric is often tied to specific political strategies. In the case of Chapter 4, 'one billion items of PPE' was used to distract from the mishandling of PPE during the early part of the pandemic. In contrast, discourse is less instrumental. It refers to the broad structures of meaning. In doing so, it emphasizes how individuals give meaning to something through a collective and relational process of understanding and talking about the world around us. This process can be structured by powerful elites, but their vision for how something should be conceived is not mapped perfectly onto reality.

Crucially, we need to see how these types of communication sit in relation to each other. Rhetoric and discourse can be placed on a spectrum of meaning, stretching from the specific to the broad.[4] The political rhetoric of the one billion items of PPE began with the articulation of the number by politicians in a press conference. But then this number entered into public discourse. To understand how such a number represents 'hugeness' meant drawing on the history of the transition from Roman numerals to Arabic numerals. This, in turn, helped understand what work this specific rhetorical strategy of 'hugeness' was doing in the press conferences.

Chapter 2, on the other hand, focused on the data bounds of *Trade-Off* and *Protect Both* – looking at how macroeconomic and public health data combined with government policies during the pandemic to create broad structures of meanings. Despite the focus on the broad, it used Devi Sridhar's interview on Channel 4 as a 'way in' to these macro-level discussions. Here, the specific provided a specific articulation of the broad.

These examples emphasize the need to reject an *either-or* choice of the specific or the general when engaging with numbers and meaning. This should be replaced by a decision of which *end of this spectrum* researchers want to either start from or pay the most attention to. Both ends offer a way into the communication of numbers. But moving from one end to the other and back again is vital for any researcher wanting to delve into quantitative meaning.

This spectrum of communication is set within different types of media. It is this media that allows communication to extend beyond face-to-face interactions. We can see how media and communication come together below:

- **Television:** Chapter 4 refers to the political communication by politicians at daily press briefings.
- **Digital videos:** Chapter 7 centres on a broadcast on YouTube by Independent SAGE and Chapter 6 uses a YouTube video by Sky News.
- **Digital texts:** Chapter 3 traced the history of 'close contact' by starting with a news article and going through public health texts, academic articles and books (including digitalized texts).
- **Social media:** Chapter 2 uses a social media post by Channel 4 as its case study.
- **Printed texts:** Chapter 7 refers to newspapers and Chapter 3 refers to academic texts printed pre-digitalization.

As with the type of communication, these media do not exist independently of each other. At the level of production, we can see how printed texts are almost entirely produced through digital processes. To design and print a poster is to engage with computer software. At the level of use, we can see how multiple types of media intersect. For example, the members of Independent SAGE provided weekly *digital videos* on YouTube, advertised these sessions on *social media* and linked through to *digital texts* from the scientific community during their presentations.

So, we have a spectrum of communication – ranging from the specific to the general – and different types of interconnected media. Up to this point, however, there has been no discussion of those communicating within this media ecosystem. The focus of this book has been upon three broad groups:

- **Expertise:** Scientists, researchers, government experts set within universities, think tanks and the civil service (Chapter 3 and Chapter 7).
- **News media:** Journalists, opinion piece contributors and editors existing with news organizations and press agencies (Chapter 5 and Chapter 6).
- **Politics:** Politicians and civil servants operating within political parties, civil service bodies and the broader political system (Chapter 2 and Chapter 4).[5]

As with the forms of communication and different media, these groups were not isolated. We can think of the way journalists interview experts to respond to politicians' announcements about public health policy. Or how politicians draw on the latest scientific research to push against narratives from the news media.

Brought together, we conceptualize numbers as circulating within a media ecosystem that involves different (a) types of communication, (b) media and (c) groups of communicators. It is this ecosystem that allows the quantitative – statistics, league tables, metrics, indicators, projections, and so on – to form broad structures of meaning around specific topics. In the case of this book, it was these media ecosystems that facilitated the creation and establishment of data bounds of the pandemic. Without the media ecosystem, the quantitative would struggle to become meaningful in the way it did.

Interrogate and appreciate quantitative realism

Underpinning this meaning was also quantitative realism. Put simply, this means believing in the ability of numbers to reveal something of the world around us. Crucially, these numbers are seen as a *better* way to reveal reality compared to qualitative information. For example, the number of apples in a bowl is given an objective certainty that the taste or smell of the same apples are not.

But this belies a more nuanced truth. To quantify something is to define and categorize an immensely complex world. This means squashing reality into conceptual boxes that do not quite fit. Using these conceptual boxes, the data then needs to be collected. This means 'finding' the phenomenon that needs quantifying and then measuring it, counting it, weighing it and so on. This turns reality into a dataset. This dataset is then subjected to different types of analyses, ranging from simple counts (for example, daily case numbers) through to complex modelling of the reproduction number (R).[6] This means numbers are a representation of the world, not objective descriptors of it.

To represent the world reliably and accurately with numbers is clearly a technical exercise that requires a high level of skill, long years of experience and unquestionable expertise. We can see this in the multiple ways that deaths could be counted during the pandemic: where coronavirus was listed on death certificates, 28 days after a positive test and all-cause excess mortality. Each measure has its disadvantages and advantages, as discussed in Chapter 2, and forms an active and critical discussion between statisticians, actuaries and data scientists. But this book emphasizes how producing quantitative knowledge is more than a technical pursuit by well-trained experts. This means we need to interrogate and appreciate quantitative realism in equal measure. To do so, we can return to the triad of language, measurement and documentation.

Language, measurement and documentation

Language, measurement and documentation sit at the heart of quantitative realism. This is best illustrated in Chapter 3 with the history of '15 minutes' and 'two metres' – the parameters of 'close contact'. This one-size-fits-all approach of *inside two metres for 15 minutes or longer* and *not inside two metres for 15 minutes or longer* presented the phenomenon of transmission as very certain. But this was misleading.

Despite being presented as scientific knowledge, the 15-minute component of the definition emerged from a common-sense science of risk. This time parameter felt about right from the existing evidence and understanding of coronavirus. The two metres distance, on the other hand, was part of a broader consensus that coronavirus spread through large droplets that travelled a short distance from an infectious person. Even though this science of transmission was increasingly disputed, the two metre demarcation of distance remained.

Chapter 3 argues that these two numbers (and other types of quantification) could confer such certainty through an intersection of linguistics, measurement and documentation. The universal language of mathematics allowed for distances and times to be presented as objective properties of the world around us. This idea was materialized by measurement devices. Highly accurate tape measures and clocks allowed for the language of numbers to be put into practice, meaning scientists could measure space and time objectively across different eras and in different parts of the world. These measurements were then recorded in scientific documents, combining the language of numbers (for example, 1, 2, 3), other languages (for example, English) and visual representations (for example, graphs). Whether lab notebooks or published journal articles, the documents themselves had the power to turn a messy, complex and subjective process of scientific research into a fixed text that was ordered, simple and objective. From this text, the number often emerged as the most iconic piece of knowledge.

But the significance of language, measurement and documentation stretches beyond contemporary sciences. Chapter 4 uses the case study of 'one billion items of PPE' to dive into the history of Arabic numerals in Europe. With their introduction into Europe from the 13th to 16th century, they changed the way we quantify the world around us. This was both a change in linguistics and documentation: the continent shifted from Roman numerals and counter boards to Arabic numerals and inscription. In doing so, there was a distinct change in the notion of scale. Europe could now deal in millions, billions and trillions.

But understanding such scale sits at odds with human experience. While it is easier to write out one billion (1,000,000,000), it is very hard to actually conceive what this number describes. Therefore, these large figures often function to convey 'hugeness' – a mathematically rooted notion of scale that involves conceiving of an abstracted notion of 'one billion' rather than an absolute one. In this way, the language of numbers is both absolute and abstracted.

This absolute-abstract 'hugeness' holds considerable power in public communication. It operated to erase the failings of the government in managing the supply and distribution of PPE by emphasizing the scale of the amount of PPE they had secured. And it hid a deeper truth: that the mathematical basis for 'one billion items of PPE' was bogus.

The tail that wags the dog

This points to an important lesson from Chapter 4: how the need for data bounds often underpins a desire to believe (and make others believe) in quantitative realism. This can often be regardless of the *actual* quantitative realism that these numbers, statistics, metrics and tables have. In the case of 'one billion items of PPE', the need to emphasize the success of government policy meant that a figure needed to be fabricated and presented as a truth. In other words, it can be the tail that wags the dog.

This sits at the heart of Chapter 5. It argues that journalists needed to construct the infodemic to reassert their credibility in the face of decreasing trust in the news media. Journalists positioned themselves as the trusted gatekeepers of information that can educate the public – protecting them from the harms of misinformation circulating online. The need for this data bound meant that journalists uncritically accepted numbers that supported their position. This was despite a wealth of empirical evidence that the high amount of misinformation circulating online during the early part of the pandemic was not resulting in mass 'mis-behaviour' in the UK. In fact, there were high levels of adherence to public health measures.

Despite this, the news media actively constructed the infodemic as a phenomenon that was underpinned by robust quantitative information. They positioned statistics as reliable representations of people's belief in misinformation, instead of recognizing the problems of using simplistic survey designs to make definitive claims about the human disposition. Journalists over-reached in their use of data, an over-reach driven by a need for quantitative realism. Once this was established, they created a causal link between people's beliefs and their actions (or future actions).

This specific story is not an outlier. A wealth of empirical research has emphasized how the quantitative plays a role in establishing the professional boundaries of journalism (Roeh and Feldman, 1984; Nguyen and Lugo-Ocando, 2016; Van Witsen, 2018) and works to establish 'truth' and 'objectivity' in politics (Coleman, 2018; Lawson and Lovatt, 2020). So, we can see how the belief in quantitative realism is something that is both *needed* and *produced*.

But how does quantitative realism differ from data bounds? We can see quantitative realism as a belief in the ability of numbers to represent the world. This belief underpins the formation of data bounds. But data bounds are specific contexts where data becomes a meaningful way to understand, engage with, discuss and experience highly quantified phenomena. These can be broad like *Trade-Off* and *Protect Both* – or specific – like the infodemic. Either way, they have the same thing at their heart: the contestation of power.

Examine how data bounds can maintain or challenge power

When we think of 'data bounds' and 'power', most scholars will begin from a negative standpoint. They point to the way data restricts, curtails and oppresses. But data bounds can also be emancipatory, liberatory and solidarity-forming. This encourages us to examine the purpose of data bounds. Their public health imperative can solidify the power of the state, their ethical function can challenge policy and suggest alternatives and they can be political requirements to deflect criticism and maintain existing power structures.

Public health imperative

In the conventional understanding of data in academia, power is exerted each time someone or something demands that a phenomenon be quantified. This can be found in the Paris lectures by Foucault (2007) from the late 1970s. He argues that quantification makes something 'known' so it can be governed. To determine 'the crime rate' is to manage criminal activity by introducing policies that decrease this rate. In this view, to create data bounds is to manage and control the phenomena quantified.

Others focus on the way digital quantification exerts power. Here online interactions, moments and transactions are turned into big data that can be packaged and sold to companies looking to sell products and services to people and companies. This is what Zuboff (2019) calls 'surveillance capitalism'. In this view, to create data bounds is to make money from it.

Both of these positions underpin much of the writing about the pandemic. Both private technology companies and governments have come under heavy criticism for their creeping surveillance practices. This is best captured in the development of contact tracing applications, where pre-existing surveillance by technology giants Google and Apple was utilized by nation-states in tracking the interactions of confirmed and suspected people with coronavirus. For many within 'data justice' – a branch of Critical Data Studies – this represented too much unregulated surveillance of citizens (Taylor et al, 2020).

This book echoes some of this sentiment. The vast banks of public health and macroeconomic data were clearly on show in Chapter 2, while Chapter 3 pointed to the prevalence of the government's Track and Trace mobile phone application (reaching around 30 million downloads by the end of 2021 (Statista, 2022)). The scope of this disease surveillance did allow the government to 'know' (if imperfectly) certain things: the rate of infection, changes in transmissibility, average number of contacts an infected person had, the effect of lockdowns on purchasing behaviour, and so on. This knowledge allowed them to introduce policy and manage these metrics.

But this book does not start from the position that this process is inherently 'bad'. Instead, it sees data as an essential component of an effective disease surveillance system – a system that can save lives and livelihoods when matched with scientifically informed policy (aligning with the work of Bratton, 2021). Within *Trade-Off*, the government could see when cases would place hospitals under severe pressure – allowing them to time national lockdowns effectively. Equally, when they saw that the reproduction rate of the virus was acceptably low, they could look to open up society again.

In countries that had a better disease surveillance system matched with elimination and containment strategies, data about cases allowed them to effectively suppress the circulation of coronavirus. This is not to say there are no ethical issues in the level of disease surveillance. Rather, it is to recognize that effective disease surveillance does not automatically equate to morally repugnant governance of people. In other words, data alone is not objectionable. It can bolster the power of the state, while simultaneously protecting its citizens.

Ethical purpose

Data bounds do not just serve to protect citizens from a public health threat. They can serve an ethical purpose that empowers citizens by challenging power structures. We can observe two main challenges to power during the pandemic.

The first occurred inside *Trade-Off*. Here the need for a trade-off between health and economics was accepted as the terms of the game. But the nature and timing of the trade-off were challenged. The most obvious example comes from the start of the pandemic. It is well documented that the government was told by their science advisors in late February about the ways that infections could be reduced (what later became known as 'lockdown'). This advice was given after a wave of lockdowns occurring in Asia and during the introduction of measures across the European continent. Despite this, the Conservatives decided to lockdown just under one month later, on 23 March 2020. This delay cost lives.[7]

But more fundamental critiques emerge from those who recognized that *Trade-Off* was one of many data bounds for understanding the pandemic. Most notably, *Protect Both* provided a space to hold power to account by pointing to a viable alternative in the way politicians, experts and journalists understood and responded to COVID-19.

A critical interpretation of the data points to a clear distinction between countries that took different long-term approaches to the pandemic. England experienced a high number of cases, hospitalizations and deaths *and* poor macroeconomic outcomes during the pandemic. This was not particularly out of step with certain Western European countries that took a similar approach to England of moving from lockdowns to opening up society to lockdowns, and so on (what is termed a 'mitigation' approach). But it stands in stark opposition to those countries that adopted an elimination or containment strategy. So, these poor health and economic metrics can be put down to an overarching policy of mitigation and the delaying of key decisions to implement restrictions (Wu et al, 2021; Oliu-Barton et al, 2022).[8] In thinking within *Protect Both*, a Pandora's box of critiques emerge. The problems of the pandemic can be rooted in mistakes and ideologies that preceded it.

The threat of a global pandemic was well known to politicians, experts and journalists (Cheng et al, 2007; Lu et al, 2020; Sridhar, 2020). And this was reflected in the level of official planning taken by UK authorities. Before the pandemic, it was ranked alongside the USA as one of the best prepared nations for a global disease outbreak (NTI and Johns Hopkins, 2019). But it seems that there were two problems: the UK was planning for an influenza outbreak (not coronavirus) and the planning had very little substance.[9]

A notable pandemic risk scenario was carried out in 2016 called *Operation Cygnus*. This dry-run showed that 'Britain would be quickly overwhelmed by a severe outbreak amid a shortage of critical care beds, morgue capacity and personal protective equipment (PPE)' (Gardner and Nuki, 2020). Instead of using this to spur action, there was more political will to shelve the report. In part, this can be explained by the dominant political ideologies from the late 1970s onwards. The dominance of neoliberalism over the past 40 years, and the recent obsession with austerity (2010 onwards), meant that the incumbent government had neither the actual state apparatus, the imagination or the will to prepare and implement necessary big state intervention (Errington, 2020; Whitman, 2022). While there are some success stories in how the state wielded its apparatus – such as the furlough scheme – there are many more failures (Health and Social Care Committee, 2021). Most notably, that Track and Trace was still not functional a year after it was set up (Indie SAGE, 2021).

This critical narrative of the pandemic allows us to identify a political catastrophe. The high number of cases (and resulting hospitalizations and deaths) matched with poor macroeconomic performance was a result of delayed decisions to lockdown, a poor overarching strategy to deal with the virus and long-term structural problems in the UK. Ultimately, the government failed to protect its people from a public health risk and economic hardship.

Thinking *within* and *outside* data bounds to challenge power is not unique to the pandemic. It sits at the heart of data activism and data justice (Crooks and Currie, 2021, p 204). There is the long-term project of academic organizations, such as Global Data Justice (GDJ), and political activism – think of Data for Black Lives (DFBL). These share the premise that data often supports power but it can also be used to challenge and make changes for good.

For example, we can see how crime and justice have been structured by the creep of data infrastructure. Much of the conversation now centres on reducing crime, conviction and re-offending rates through the automatic identification of criminal behaviour or the dystopian push for pre-emptive action to stop crimes from occurring. Those challenging power identify the problems of this data bound by pushing against a simplistic link between technology and justice. Some point to the way automatic surveillance technologies misidentify non-white people more often, leading to more wrongful arrests of these groups (Johnson, 2022). Others emphasize the way pre-emptive policing will proactively change society through data-veillance (Arrigo and Sellers, 2021).

Despite these efforts, the ethical potential of data bounds is often superseded by the power of data bounds to maintain the status quo. Returning to the pandemic, we can see how *Trade-Off* helped deflect criticism from the government and maintain the political establishment.

Political requirement

In face of criticism, there was a political requirement of *Trade-Off* to do two things. First, it needed to foreclose space beyond *Trade-Off* by discrediting or ignoring *Protect Both*. This was done through a combination of explanations about why the UK could not introduce elimination or containment strategies: we had a political system that was not as authoritarian as China, we had a culture that had less appetite for the restrictions it would place on travel, we were a financial and transport hub, to name but three. However bogus these reasons, they were effective at relegating *Protect Both* to a fantasy that exists in China, New Zealand and Singapore – but could never cross the border into the UK.

Second, it needed to position political decision and long-term structural changes as working to render the government's handling within *Trade-Off* as a successful one. This was done through a constant denial of mistakes – the government did not delay the first lockdown too long, the government did not encourage too many infections with their 'eat out to help out' scheme, the government was not providing contradictory and confusing public health advice, and so on.

This book is testament to the achievement of both positions. The pandemic failed to make any real dent politically. While survey findings might point to disapproval of government decisions, this has not resulted in a substantial change to the political system from the start to the so-called 'end' of the pandemic. The Conservatives who held power in early 2020 were still at the helm. They had the same leader – Boris Johnson – that won the 2019 General Election. And after the necessary state spending in the early stages of the pandemic to preserve the current political-economic structure (Blakeley, 2020), limited government expenditure returned as the orthodoxy of conservative government. Some may rebut this conclusion as too totalizing and blunt, maybe pointing to the government's raising of national insurance contributions to fund social care reforms.

But if this is the result of a catastrophic pandemic, those rebuttals would be operating within an extremely narrow scope of what counts as change. The inability of the pandemic to lead to such changes also means that data does not hold the power that many ascribe to it (a point made in Balani, 2020). In a hopeful view, data bounds are supposed to reveal an unjust reality so it can be changed. But what could be seen as more unjust than such a large-scale loss of life, high numbers of long COVID, delayed hospital care, entrenchment of health inequalities, missed education and lost livelihoods? At a time where society could use data to 'sense itself' through numbers (Bratton, 2021), it was left numb to a political catastrophe. To paraphrase Crooks and Currie (2021), more data is not the answer. Data did not reveal injustice, it obscured it. Data did not push for alternatives, it supported poor policy decisions. Data did not help us understand, it aided ignorance.

In an avalanche of meaningful numbers about society, it seems that these numbers failed to crash down on dominant political institutions. Despite pointing to very serious failings of government, they could not form part of a broader push to change political leaders and parties and/or reform the current political system. Given this, it becomes vital to understand how these types of data bounds dominate and others do not. If all data bounds are set within the media ecosystem, have quantitative realism at their heart and function to challenge or maintain power, what sets them apart? This question sits at the heart of the final part of the toolkit.

Determine why some data bounds dominate over others

Determining why some data bounds dominate and others are marginalized is immensely difficult. We can begin by acknowledging that there is little certain in this process. There is nothing inevitable about media ecosystems, quantitative realism or the challenging or maintaining of power. But this should not push us into a conclusion that says all data bounds have an equal chance of occurring. *Trade-Off* helps us tease out some of the factors that allow certain data bounds to succeed. It is a particularly good example for two reasons: it was established very quickly and remained extremely strong in the face of alternatives.

Political consensus

The introduction above said that there was nothing inevitable about a data bound flourishing because it challenged or maintained power. But this is a bit misleading. The lesson from this book is that a data bound that shores up powerful institutions is more likely to dominate over those that do not. In other words, if more people in powerful positions agree to the terms of a data bound, the more likely that said data bound occupies a commanding position in society. In the context of the pandemic, this can be called a 'political consensus' – a broad agreement about how politics should respond to the virus.

We can distinguish between two elements of this consensus. The first concerns actual government policy. The Conservative government adopted a mitigation approach that looked to manage COVID-19 by moving from national lockdowns to opening up society to national lockdowns. The timing of each intervention depended (at least notionally) on data: rising cases would trigger a lockdown, plummeting GDP figures would lead to opening up society, rising cases would trigger a lockdown, and so on. So, we can see the mutually reinforcing relationship: the constant back-and-forth between situation 1 (poor health indicators, good economic figures) and situation 2 (good health indicators, poor economic figures) cemented the idea that *Trade-Off* was logical. Each time the UK implemented policy to move the country from situation 1 to situation 2, the inexorable trade-off between health and the economy was cemented.[10]

But the dominance of *Trade-Off* is not just about the policy of the incumbent government. These policies were set within a broad political consensus. In an alternative universe, the opposition parties would argue for the implementation of an alternative strategy for dealing with COVID-19. This did occur in some countries that took an elimination approach. Most notably, New Zealand saw a split between the ruling party and the opposition, who argued that elimination was too extreme and draconian as a government policy. While the data bound of *Protect Both* dominated, there was scope for imagining alternative data bounds that were not based on reaching and maintaining 'zero COVID'.

But this plurality was largely absent in the UK. There was a broad political consensus in support of the government's approach. This was most clear when we consider the main political parties and media organizations. The two main opposition parties to the Conservatives – the Labour Party and the Liberal Democrats – and the majority of news organizations – including The Telegraph Media Group, Daily Mail and General Trust, Reach Plc, News Corp UK and Ireland, Sky Group – accepted the premise of *Trade-Off*. When they did provide a critique of the government, it operated within *Trade-Off*: they pointed to the lateness of government intervention or highlighted the failure of the Track and Trace system to actually mitigate the effects of the virus.

This broad consensus regarding the political response to coronavirus structured the media ecosystem. We can use Hallin's (1986) classic concept of 'spheres of legitimacy' as a useful heuristic here. At its centre is the *sphere of consensus*, where the main individuals and organizations agree. In our case, this would be the data bound of *Trade-Off*. The second is the *sphere of legitimate controversy*, where the same individuals and organizations have accepted disagreements – for example, the timing of lockdowns within the broader acceptance of *Trade-Off*. The third is the *sphere of deviance*, where *Protect Both* existed. The lack of news media organizations, political parties and other prominent or powerful individuals voicing their support for *Protect Both*, relegated this data bound to the fringes of public discourse.

Political consensus (a consensus about the role of politics by those inside and outside Westminster) makes a data bound more likely. The stronger the consensus, the stronger the data bound. But this consensus is born from and into a history. It was not just something that emerged during the pandemic – set apart from the contingencies of space and time. Therefore, we need to consider the importance of historical context in how data bounds come to dominate.

Historical context

Chapter 7 approached the role of history through the lens of 'historical norms'. At its broadest level, this means interrogating what is considered normal in a society at a specific time – and, in turn, how this notion of normal is not independent of the history of said society. To explain how normalization relates to data bounds, Chapter 7 tied together the history of inequality with the data bound of *Trade-Off*.

In early July 2021, Christina Pagel projected that there would be 90,000 cases per day by the middle of the same month. Despite being a well-respected scientist who made regular appearances in the media – this figure failed to circulate. Some could argue that it was due to its crude nature – and how it ultimately proved to be quite far from the mark. But this explanation is too quick to ignore the poor projections that *did* circulate.

Instead, the chapter argues that the threat of this projection was not considered substantial enough because it would disproportionately affect the poorest in society. Instead of breeding society-wide anger and will for change, there was an accepted logic: yes, 90,000 cases a day is a lot – but these higher cases, hospitalizations, deaths, long COVID, missed school days and missed workdays would only substantially affect the poorest in society. The acceptance of such inequality was due to the normalization of inequality in Britain.

As with all historical accounts, the chapter struggles to do justice to the myriad historical processes that culminated in the normalization of inequality. But it looks to key events in the history of the UK, focusing on the role of neoliberal policies and austerity in widening inequality over the past 45 years. This has fostered a collective view – from politics, to experts, to media organizations – that gaps between the rich and poor are normal. Such an argument becomes more convincing when we compare the UK to other countries that *did not* take these political choices. In these contexts, we have seen inequality narrow.[11]

Normalization is just one way to place data bounds within a historical context. Another, rather more popular approach, is a Foucault-styled 'genealogical' appraisal. This involves an emphasis on 'a form of history which can account for the constitution of knowledges, discourses, domains of objects etc.' (Foucault, 1980, p 117). Such a focus is – like much of Foucault's proposals – better understood when put into practice. A good example of which is C. W. Anderson's (2018) genealogical account of the quantitative in journalism. His book *Apostles of Certainty* places the contemporary fascination with 'data journalism' as a phenomenon that is baked in a distinct North American history. Whether scholars choose to take this type of historical approach or not, the message is clear: data bounds cannot be divorced from the contingencies within which they emerge.

Experience

Returning to the nature of data bounds themselves, we can see the significance of experience. For data to become a meaningful way to engage with a phenomenon, it needs to relate – in some way or another – to people's lives. Some data bounds are more effective at doing this than others. As this book has highlighted, *Trade-Off* was particularly strong because it directly related to people's experience. It represented a collective memory of the waves of the pandemic and provided a space for people to make sense of their current practices, emotions, relationships, thoughts and actions during the pandemic. The idea that the government needed to 'open up' or 'close down' the country depending on diametrically opposed health and economic indicators acted as an explanatory framework for living within the pandemic – one that was reinforced each time the government decided to 'open up' or 'close down'.

The distinct 'humped' graph of cases, hospitalizations and deaths was the most complete single representation of *Trade-Off*. It featured heavily in how politicians explained the pandemic to the public. It was the image that was referred to in March 2020 as the rationale for introducing a national lockdown, immortalized in phrases such as 'flatten the curve' and 'squash the sombrero'. It was the same image, updated with the latest public health data, that was presented in June 2020 as the reason to enjoy our summers with fewer restrictions. And it was this graph that forewarned of a lockdown in late 2020.

But it did more than legitimize government policy. It tapped into a collective experience of the pandemic – rising cases matching with implementations of lockdowns, decreasing cases corresponding to the opening up of society, the troughs lulling us into an idea that the worst may be over, and the peaks capturing the feeling that it may never be. This experience was characterized by a complex web of emotion: a constant low hum of stress, combined with acute periods of shock, anger and sadness. Understanding such an experience is inevitably complex and open-ended. Chapter 6 provides one way to engage with it: taking a genre studies approach to tease out the relationship between emotion, experience and performance.

It argues that the affective power of the graph was rooted in its ability to represent the *uncertainty* of lurching from locking down, to opening up, to locking down but also the *tragedy* of having over 1,000 deaths per day due to the virus. When presenting this graph, journalists had to match their performance with the sombre nature of the visualization. When Roland Manthorpe broke this unwritten rule in his piece for Sky News, the violation of the taboo could be *felt* by the viewer.

But inferring that people experience data from a textual analysis is not water-tight by any means. So, we can turn to other academic work exploring the role of experience and emotion. We can point to the writing of Kennedy and Hill (2018) on emotion, cited at the end of Chapter 6, the emphasis on feeling in Woodward's (2009) writing on 'statistical panic' and 'statistical boredom' and the importance of narrative for people and communities to relay and understand data (Crooks, 2017; Dourish and Gómez Cruz, 2018).

This disparate set of work is brought together expertly in the work of Crooks and Currie (2021). Rooted in their own experiences of working alongside working-class black and Latinx activists in Southern California, they reject a rationalist approach to data for a view of data that 'motivates people to act on their passions and imaginations'. The quantitative is both affective – it elicits emotions – and narrative – it is tied into and around stories. The slogan of 'We are the 99 percent!' from the Occupy movement provides the most well-known example of this: it captures the emotive injustice of inequality between most of us (99 per cent) and a select few (1 per cent) but also constructs a narrative about how we are consistently exploited, what facilitates this and how it can be challenged (Crooks and Currie, 2021). In other words, numbers are experienced through emotion and narrative – not just the classic rationalism that often frames how we think of the quantitative meeting the public. Importantly, as the work of Crooks and Currie has shown, this experience is best understood through ethnography.[12]

A runaway train of meaning

Whereas the previous explanations have encouraged the scholar to look beyond a strict focus on the quantitative, the fourth returns to the data itself. It focuses on the haphazard and chance-laden way certain data bounds take centre stage. It emphasizes how the quantitative can be thrust into the media ecosystem and – due to factors outside the control or expectations of those placing it there – can become a runaway train of meaning. Or, in the words of Genevieve Bell (2018), how data can go 'feral' in the 'wild'.

In Chapter 6, we can see how such a process occurred with the counting of deaths from COVID-19. This number was needed as part of the government's disease surveillance system, so they could effectively introduce public health measures within *Trade-Off*. The data concerning deaths was produced by government and quasi-government organizations (for example, the NHS and the ONS) and then communicated by these organizations and politicians at press conferences through text, speech and visualizations. The visualization of peaks and troughs of daily death counts came to represent the ultimate toll of the pandemic on people's lives. The graph was constantly updated with the latest daily deaths and adjustments from changes in definitions. The graph then became an ever-updating icon that journalists had to interact with and explain. The image became further entrenched in society. When Roland Manthorpe came to present the graph in November 2020, the visualization had become so iconic that his overenthusiastic reporting sat at odds with the emotion attached to the graph.

The meaning attached to this visualization by November 2020 is hard to link to the original need for data from early 2020. The visualization – as it was propelled into public discourse – took on a meaning that belied expectation or control. This points to the semi-autonomous nature of the quantitative (Anderson, 2020b). To be clear, I am not arguing that the quantitative has autonomy like human beings, such as self-determination, free will and choice. Rather, that when the quantitative circulates, it can exist as a runaway train of meaning that often exists independent of the specific meaning human beings attempt to give it. The emergence of this autonomy is from the autonomy of individuals who demand, produce and circulate numbers but somewhere along the way the number itself gains a sense of power over its own life that escapes those demanding, producing and communicating it.[13]

But the ever-presence of this graph also points to a broader lesson. Because this data visualization could be constantly updated, it had a longer public life. There might be new data but the form of the data visualization remained the same: cases, hospitalizations or deaths on the *y*-axis and time on the *x*-axis.

An individual number, on the other hand, often functions as a piece of knowledge – one that can gain attention but, more often than not, is superseded by new forms of knowledge. We can see this in Chapter 5, where the life of the two figures describing belief in misinformation about coronavirus had relatively short lives. They described a certain number of views of a video and the belief in conspiracy theories at a certain moment in time. Both pieces of knowledge soon became historical.

The short life of these numerical pieces of knowledge can work in favour of those communicating them. The 'one billion items of PPE' figure from Chapter 4 seemed to play out this way. Within a week, the number had emerged, gained widespread attention and then dropped into obscurity. This all occurred before the *Panorama* documentary that uncovered the imaginative accounting involved in arriving at this large number, meaning that such an uncovering did little to foreground the number again. The uncovering was too late.

Is there any hope?

When we apply this four-part mantra, do we arrive at something hopeful? It is certainly easy to *not* be hopeful when we consider how quantitative realism can be used to prop up versions of reality that are a political requirement to protect the status quo. But to end on such a note would fail to see any positives. It is my hope – and belief – that data bounds can serve an ethical purpose.

Remaining within the confines of the six short stories from this book, we can see two possible sites where the emancipatory and progressive potential of data could be realized. First, the newly politicized scientists from biology, mathematics, physics, public health, epidemiology and other specialisms. It would have been hard for likes of David Spiegelhalter, Neil Ferguson and Christina Pagel to have imagined the attention they would receive during the pandemic. But these individuals (and others in the hard sciences) have not just been catapulted into the limelight – their relationship to politics has intensified too.

For 'the insiders' – such as Chris Whitty, Jonathan Van Tam and Patrick Vallance – they have been afforded a considerable level of power within government. This power, however, was always structured by the wider political system. They had the power to talk about the pandemic to the public, sit in on important policy meetings and organize SAGE. But they could not push for considerable change (at least not publicly). And this is how they maintained their power. Because they did not criticize publicly, they stayed on the inside of the political system. On the other end, there were the 'outsiders' who were kept away from the government exactly because they criticized their handling of the pandemic (Cairney, 2021).

One of these, Christina Pagel, features heavily in Chapter 7. A Professor of Operational Research in the Mathematics department of UCL, Pagel was largely unknown outside of academia before the pandemic. Being part of Independent SAGE, however, thrust her into the spotlight. This was not just because of a general push towards expertise during the crisis, it was because she was willing to use her expertise, knowledge and findings to actively criticize the government and present alternative policy options. And Christina was not the only one during the pandemic. Other members of Independent SAGE, such as Kit Yates and Deenan Pillay, have played similar roles to Pagel: drawing on the 'hard sciences' to form political arguments.

But Pagel, and her colleagues who were both political and critical, remained as outsiders. They formed well-researched arguments that did often circulate in public discourse yet had very little effect on changing the government's policy.[14] So, these outsiders were largely restricted to a discursive role rather than a policy one. Sadly, hope does not seem to lie in these experts – at least not under the current political system.

But what about the data journalists? The work of the *Financial Times* and *The Economist* – publications adept at using and reporting numbers before the pandemic – have spearheaded much of the most impressive data reporting during the pandemic. The pandemic seems to have pushed these organizations, and others, to take data journalism into the world of data science. That is, to move away from the presentation and interpretation of open data sources and towards the creation of data and the adoption of complex data analysis. To outline the hope in this change, let's look at two examples.

The first comes from Chapter 4. When mapping the different data bounds, an infographic from the *Financial Times* was cited as early evidence for *Trade-Off* being misleading. By plotting the change in GDP on one axis and cumulative deaths per million on the other, and then plotting a number of countries on the scatterplot, this work went beyond merely describing the world. It put forward an argument that the current strategy of lockdown, ease of restrictions, lockdown, ease of restrictions was misguided.

The second example is one not covered in this book. In a virtual conference based at Bournemouth, three members of the team at *The Economist* – James Tozer, James Fransham and Sondre Solstad – outlined the work they had been doing with data during the pandemic. Much of their exciting work centred on filling in data black holes in the early stages of COVID-19. In the face of a lack of information about deaths in each country, they consulted unofficial sources – most notably the number of burials in Mexico. To combat the lack of data about movement, they pooled together Google Map data to 'measure people's foot movement from nearly 30 cities around the world' (Nguyen et al, 2021, p 7).

But what is all this data for? For those at *The Economist*, it often felt like more data was the goal itself rather than using that data to forge new data bounds. Whereas those at the *Financial Times* twinned data with some political argument that pushed (if implicitly) a rethink of policy. Therefore, just having more data is not the answer. It needs to be put to work. Sadly, this is an all too rare occurrence in data journalism.

So, where else can we look for hope? This book argues that the main source of hope may lie outside of the parameters of the six stories from this book. It resides in those using data – both explicitly and implicitly – in the public's activism, collective movements, campaigning and grassroots political efforts. This is where data can be infused with effective political action. These can be in organizations that have data at their core, such as DFBL covered in Kitchin's (2021a) book. This involves a direct example of data activism, where the quantitative is used to highlight problems, suggest remedies to problems, push for widespread political change and engage in community-level projects.

Alongside these types of organizations, there is a much larger set of collectives, organizations and movements that are influenced, structured and dictated by the datafication of society more broadly. Extinction Rebellion (XR) comes to mind here. While they function as an environmental group that raises public awareness and places pressure on national governments to change policies, they also engage consistently with numbers. The three stand-out figures are the *1 per cent*, the *99 per cent* and *1.5 °C*. The two percentages stand in for the rich and powerful in society (1 per cent of people) and the normal citizens (99 per cent of people).[15] It is the 1 per cent that present the greatest risk to the 1.5 °C warming of the planet. The low levels of taxes paid by the owners of Facebook, Amazon and other big multi-national companies restrict the funds for large-scale government intervention (@XRebellionUK, 2021).

Outside of these iconic numbers, there are the ways that figures serve to situate individual members of XR as part of a larger group of people. Here we can think of how marches can attract different numbers of people – those with bigger numbers often being considered more significant than others. Or how the rising or decreasing numbers of members part of XR are connected to some idea of a rising or waning political force (or the English idiom of 'safety in numbers'). At a more banal level, we can see how measurements function. During COVID-19 there was a need to observe social distancing at political protests, often meaning people kept one metre or two metres away from each other. The same political protests were organized around specific mathematical times, such as 10:15 am on 12/10/23. In other words, numbers were threaded through political protests in both the explicit and the banal.

To understand the importance of numbers in these spaces, for these causes and for these people, there also needs to be a methodology shift towards phenomenological, ethnographic and observational research. It is only through this approach that we can start to understand how people live within numbers: how people use numbers, communicate data, put statistics into practice, think about league tables, talk about projections, and so on. And it is only through this that we can see the hope in quantification.

Afterword

A lot has changed since the most recent case study from this book, taken from July 2021. After so-called 'freedom day' was introduced in mid-July, there was a summer and autumn period of fewer restrictions. The emergence of the Omicron variant in South Africa in November 2021, however, meant the Conservative government introduced a series of measures. These included red listing countries, booster jabs, advising home working and introducing mandatory mask wearing indoors.

For some, these restrictions did not go far enough – a policy closer to lockdowns was called for. A BBC News Online article explained that there were benefits of introducing a lockdown – the delaying of the peak of cases, lower pressure on hospitals, and so on – but this would also cause 'harm to jobs, mental health and education' (Triggle, 2021). While the scope of what was traded-off was expanded to include mental health and education, it still operated within the same broader paradigm of *Trade-Off*.

But the government did not introduce a lockdown and the UK did not experience comparable levels of pressure on hospital systems. It seemed that vaccine coverage enabled England to not lockdown. After the wave of Omicron reduced, so did the restrictions introduced by the government in November 2021. And as the country pushed into 2022, these restrictions have not re-emerged. This has led many to talk of a 'post-pandemic' era – one defined by optional lateral flow tests, peeling two-metre stickers on shop windows and dust slowly gathering on face masks. In this world, *Trade-Off* has become a thing of the past.[1]

For some, this would make the empirical basis of this book less relevant to the 'post-pandemic' world. But the stories traced in each of the chapters were not aimed at better understanding the pandemic per se. Rather, they pointed to six key characteristics of 'data bounds' and four imperatives for scholars looking to put this concept to work. This means that data bounds outlive the COVID-19 empirical basis upon which they are built – they should be used to understand highly quantified phenomena in the post-pandemic world.

Perhaps the most obvious of these is the global crisis of 2022: rising inflation, increasing profit margins and falling wages. Given the nature of this crisis, data bounds are particularly useful to unpick how prices, quarterly profits, average wage increases, absolute counts of job vacancies, falling trade indicators, falling tax revenues and GDP growth have become meaningful ways to engage with, think about, discuss and respond to the so-called 'cost of living crisis'. How is this quantitative knowledge established and trusted? How is it afforded power to construct reality? How is power challenged or maintained? What are the competing paradigms that are marginalized? How does the media ecosystem function? Data bounds are uniquely placed to answer these questions.

B. T. Lawson
Sheffield

Notes

Introduction

1 These two premises are heavily influenced by the work of David Beer (2016) in *Metric Power*. His three concepts of 'measurement', 'circulation' and 'possibility' underpinned much of my thinking and organizing of this book.

2 This concept has numerous academic roots. But I am specifically indebted to An Nguyen here for drawing my attention to Woodward's (2009) talk of the quantitative as 'weather', 'atmosphere' and 'emissions' and Gummo's discussion of Fred Turner's (2013) 'democratic surround'.

3 This is an iteration on Desrosières (1998) term 'metrological realism' that refers to all forms of quantification (including measuring).

4 We can compare the idea of 'quantitative realism' to that of 'capitalist realism', introduced by Mark Fisher (2009). Just as it becomes harder and harder to think *outside* capitalism, it is harder and harder to think outside *quantification*.

5 Kitchin (2021a, pp 11–12, 211) also nods to the way data lives within science and politics from 'generation to destruction'.

6 For another example of how representation has historically been linked to quantification, see Prevost's (2009) account of Fascist Italy's (1922–43) statistical weighting model that gave more 'weight' to the votes of a small number of people.

7 Added to this list is the work of *Music by Numbers* – an edited collection that has an individual number from the music industry as the focal point of individual chapters (Osborne and Laing, 2020). Also of note is the approach of Louikissas (2019) in *All Data Are Local* – she takes one example per chapter to emphasize the way data is always rooted in a specific space and time.

8 We can see a similar approach taken by Rob Kitchin (2021a) in his recent book *Data Lives*. He uses biographical accounts and fictional narratives to speak a certain truth about how we live in, through and by data in the 21st century.

9 This is what Berman and Hirschman (2018, p 265) refer to as 'mundane' forms of quantification. These 'less potent' figures are important sites of study as they highlight how certain numbers have 'little oomph' compared to other, more prominent, forms of quantification.

10 The analytical guide iterates on the scheme put forward by Simpson and Dorling (1999, pp 415–16) who refer to the four things needed to produce a 'well-known' statistic: it needs a purpose, the data needs to be assembled, it needs to be analysed and then communicated. Despite the instructive nature of this scheme, little subsequent work has referred to it. Grainger (2007) extended the framework by including three additional stages, beginning with consultations with end-users and ending with receiving feedback from end-users. As with Dorling and Simpson's original scheme, however, this approach to statistics did not really take off in academia.

11 Acts of communication are inherent to every part of the Life of a Number. But the term communication is used in a specific sense here.

12 This process of 'backgrounding' can be seen as similar to Kitchin's (2021a, pp 101–7) idea of 'the death of data'. But the focus of the fifth stage in the Life of a Number is on how there is something like a 'meaning death' rather than the deletion of data.

Chapter 2

1 This data refers to deaths by date of death (rather than when the death was reported). This relies on the 28 days after a test definition of deaths. When comparing to April, when testing capacity was directed solely on hospitals, we should see the peak as actually much higher.

2 Obviously, this comparison is not valid – given the significant lag in GDP data and the smaller delay in reporting deaths – but these two pieces of information were published and circulated on the same day.

3 See paper by Guangyu et al (2021) for detailed explanation of each strategy.

4 For a more detailed explanation of seven-day rolling averages, see Chapter 7.

5 The data used here is the government data, using a seven-day rolling average. It refers to 'cases by specimen date' rather than 'cases by reported date'.

6 This lag between cases, hospitalizations and deaths does not always follow this exact pattern of days. There is a distribution of people hospitalized and a distribution of people who died. This is merely illustrative of how they relate to each other in this period of the pandemic.

7 A longer discussion regarding the importance of local and national data collection for international comparisons (and how this locality is often erased from discussions) can be found in Louikissas's (2019) *All Data Are Local* and Jerven's (2013) book about African economic statistics.

8 When we look at death certificates, there is a separate conversation to be had concerning the difference between 'due to COVID-19' and 'involving COVID-19'. While the 'involving' approach is much broader than 'due to', it does provide an indication of the level of deaths in care homes that occurred in the UK during the early stages of the pandemic.

9 Some of these deaths would have been included in the daily count when a patient was admitted to hospital, tested and then discharged.

10 For a good summary of the promises and pitfalls of measuring social wellbeing, the work of Carr-Hill (2019a) should be consulted.

11 Other reputable analyses only use deaths for 2019 – most notably, The Institute and Faculty of Actuaries in their regular bulletins about all-cause excess mortality.

12 The *Financial Times* did later change their approach to account for population structures but, in late May 2020, they were not doing this.

13 Scotland did pursue an elimination strategy in the summer of 2020. But this was abandoned as they went into autumn 2020 due to a rising number of cases.

Chapter 3

1 This is one example of the way seemingly non-quantitative words or phrases are actually underpinned by numbers. For example, the term 'famine' used to be used by aid workers, politicians and journalists to draw attention to particular food shortages (Mamdani, 2007). In the mid-2000s, however, the Integrated Phase Classification (IPC) system was established as a standardized tool to assess food security across different countries. Under this system, a famine could only be declared if three criteria were met: (1) at least 20 per cent of households in an area face extreme food shortages with a limited ability to

cope; (2) acute malnutrition rates exceed 30 per cent; and (3) the death rate exceeds two persons per day per 10,000 persons (IPC Global Partners, 2019). In this way, 'famine' was now quantitatively defined.

2 Despite Flügge and his contemporaries finding that droplets could also travel much further than one yard. Flügge (1899) himself found that 'some droplets could be propelled as much as 4–5 feet' (Hare, 1964, pp 4–5). Randall et al (2021) also point to the work of von Weismayr and by Laschtschenko from the Flügge school. They found that plates as far as nine metres away from sneezing subjects contained cultures (Randall et al, 2021).

3 As Randall et al (2021) document, the emphasis is on airborne transmission.

4 There are a number of other papers published during this period that point to anecdotal evidence of SARS-CoV-1 travelling further than 1–2 metres. A paper by Gamage et al (2005) emphasizes the close proximity rule but also certain 'exceptions': 'The recently published studies on the hospital-associated outbreaks of SARS have all concluded that direct contact or close exposure to a SARS patient is generally required to transmit the virus, although important exceptions exist (pp 5,13–16). In some circumstances, *aerosol-generating procedures have resulted in spread beyond that which is expected by droplet transmission of less than 3 feet*' (Gamage et al, 2005, my emphasis). Added to this, a paper by Scales et al (2003) provided an 'exception' to the close contact rule: 'SARS developed in *one quarantined healthcare worker (a nurse) who had not entered the index patient's room*; the disease did not occur in any other healthcare workers who had not touched or had close contact with the index patient' (Scales et al, 2003, p 1208, my emphasis).

5 It should be noted that the UK is officially metric but still uses imperial measurements, most obviously in road distances that are still in miles and beer servings that are in pints.

6 We can see some references to '15 minutes' in earlier work on droplet transmission. In a piece of research by Downie et al (1965), they collected samples from 5–30 cm from the mouth using an impinger. These samples were collected for a period of 10–15 minutes. But this research seems to use 10–15 minutes as a 'rule of thumb' rather than with reference to specific literature. And this is reflective of all the texts that use '15 minutes' – none of them link, directly or indirectly, to an actual study on transmission that establishes 'more than 15 minutes' as the time parameter for increased risk of contracting coronavirus.

7 This explanation can be found across the scientific community. In two NPR pieces from 2020 and 2021, they quote two scientists. Emily Gurley, an epidemiologist and contact-tracing expert at the Johns Hopkins Bloomberg School of Public Health, explains that 'we don't have strong evidence for exactly what the right distance or the right duration is' (Godoy, 2020). They also quote Leana Wen, an emergency physician and public health professor at George Washington University, who explained that 'there was nothing magic about 6 feet and 15 minutes … however, we need to draw the line somewhere' (Eldred, 2021).

8 There is a different discussion to be had here regarding humans' preferences for certain numbers. In a base ten counting system, there is often a preference for multiples of 10 (10, 20, 30, 40 …) and halves (5, 10, 15, 20, 25, 30 …). For a great overview of this literature, see an article by Mitchell about number clustering in the financial markets (Mitchell, 2001).

9 This definition did undergo some alteration though. Before January 2021, the 15 minutes spent within two metres of someone was approached as a full block of 15 minutes or more (for example, going to a café with someone, sitting opposite them on a small table for 25 minutes would be included). After January 2021, these 15 minutes could be made up of several shorter interactions over the space of a day. For example, if you were within two metres of someone for five minutes in the morning, five minutes at lunch and five minutes in the evening (Donnelly et al, 2021).

10 The two-metre rule was officially changed to one-metre-plus in the summer of 2020. As this was largely to allow hospitality to have a larger capacity, the idea of one metre never really stuck.

11 For a more detailed discussion about measurements, there is a range of fascinating books. You can consult Arnold's *Building in Egypt: Pharaonic Stone Masonry* (1991) to appreciate the importance of the Egyptian cubit rod – a wooden or stone device that provided standardized measurements of distance. If you are more interested in the measurement of time, Landes (1983) has written a great book on mechanical clocks and Porter (1995, pp 22–3) maps the shift to standardized time in Medieval Europe.

12 There is a broader discussion that can be had here regarding the way that scientific work is legitimized by the scientific community to create accepted scientific knowledge.

Chapter 4

1 It reared its head again a month later when the government announced that it had signed a contract for two billion items of PPE (26 May 2020).

2 One billion here refers to the 'short scale' billion of 1,000,000,000 or 'one thousand million'. It was only in 1974 that Prime Minister Harold Wilson confirmed that the UK government would only use 'one billion' in the short scale. Before this, both the short scale and the 'long scale' (one million million or 1,000,000,000,000) were used.

3 It should be emphasized that the use of Roman numerals from their conception through late Medieval Europe went through many iterations – they were not a fixed and standardized mathematical language as we are used to today.

4 A contemporary account of the Dunbar Number is outlined in *Scale* (West, 2017).

5 There are additional problems that have not been documented here, namely the way PPE could not be used because it failed to meet relevant standards.

6 There is a wider issue that will not be addressed in this chapter regarding the huge impact of austerity and privatization on the ability of the government to respond to the pandemic.

7 Buckland was quoted as saying, 'We needed to make a choice about testing, we did decide to focus upon the NHS [over social care]'.

8 A discussion about the different ways of counting deaths, including an explanation of 'involving COVID-19', is to be found in Chapter 2.

9 Research conducted by Cambridge University Hospitals NHS Foundation Trust later found that the wearing of a high grade mask (known as FFP3) can provide 100 per cent protection from the virus (Shukman, 2021).

10 This stands at odds with the existing literature on rhetoric and statistics that emphasizes the way individuals adopt rhetorical strategies in their use of numbers, from prosopopoeia to alliteration (McCloskey, 1987; John, 1992; Kilyeni, 2013; Katchergin, 2015; Avilés, 2016; Koetsenruijter, 2018). This literature pays attention to how numbers are *used* rather than the properties of the numbers themselves.

Chapter 5

1 This explanation refers specifically to the widespread behaviour during the early stages of the pandemic (between March and July 2020). I do not dispute that 'mis-behaviour' has existed, to varying degrees, across the pandemic as a whole.

2 This is called an 'infodemic moral panic' by Carlson (2018). While this term is useful, this chapter attempts to use elements of Carlson's work without the jargony label of 'moral panic'.

3 While this chapter does not focus on the notion of 'representation' here, it is worth outlining a working definition for those interested in looking into this further. Chouliaraki

(2010) conceptualizes representations as symbolic power that 'coexists with and reproduces, but may also change, dominant relationships of power (economic, political and cultural).

4 While the number resurfaced in US news media discourse in August, with the emergence of the second instalment of *The Plandemic*, it was only mentioned sporadically after May 2020 in news media content in the UK.

5 This can be positioned within the broader mathematization of the sciences covered in Chapter 3.

6 For an account of this history, you can consult the following texts: Moon (1999), Worcester (1991), Holli (2002) and Fried (2011).

7 A peer-reviewed academic article was published alongside Ipsos Mori's non-peer-reviewed report in June (Allington et al, 2020).

8 There is extensive literature on the 'do's' and 'don'ts' of surveys, including a large amount of work that takes a critical approach to surveys. These can be found in survey handbooks, such as the comprehensive *SAGE Handbook of Survey Methodology* (Wolf et al, 2016). There are also specific chapters and journals that are useful touchstones:

- Evans (2019) provides a commentary on adult skills surveys, paying particular attention to conceptualization of variables, selection of samples, difference between correlation and causation and commensurability.
- Carr-Hill (2019b) emphasizes the way that surveys are often poor at estimating poverty, largely because of the way certain populations are omitted by design and others are under-represented in practice.
- Randall, Coast and Leone (2011) provide an excellent account of the conceptual rigidity of 'household' in household surveys, a limitation that negatively affects the quality of the data from household surveys conducted in the Global South.

These are very useful resources that should be consulted by an interested reader. The following section attempts to tease out these criticisms through a set of fictional accounts of how people might interact with surveys.

9 For a more specific account of how data is collected and processed on social media, see Tear and Southall (2019).

10 A similar argument can be made for surveys. If they are used to capture some essence of humanity (desires, beliefs or motivations), they are considerably limited. But if they are used to refer to more instrumental ideas, such as where someone goes most often for their news, they can prove useful as broad (and flawed) overviews of a phenomenon.

11 It is important to stress that this chapter looks at how misinformation is communicated by journalists, not the phenomenon of misinformation itself. I currently work on a project called *The Everyday Misinformation Project* as a research associate, alongside Andrew Chadwick, Cristian Vaccari and Natalie-Anne Hall. This research looks to place misinformation within people's experiences of using personal messaging applications. It starts from the point that misinformation can be harmful (for example, in people's views of vaccinations), but emphasizes the complexity of misinformation. Part of this, is to see the effects as complex, that is, not automatically associating exposure to misinformation as leading to 'mis-behaviour' or a lack of trust.

12 For an early analysis of the cost of this late lockdown, see Annan's (2020) simple model that estimated 30,000 lives were lost in not locking down on 16 March 2020.

13 There are two key points to be made here:

- There are great examples from journalists that are not covered in this chapter. Roland Manthorpe from Sky News has done an excellent job of tempering the hysteria around misinformation and lockdown rule breaking. Analysis like his is sorely needed.

- The shifting tide of what is classed as 'conspiracy theory' or 'legitimate explanation' is an unpredictable one. One example comes from the coronavirus origin theories. For much of 2020, the idea that the virus leaked from a Chinese laboratory in Wuhan was considered wild conjecture – mainly confined to Donald Trump and the conspiracy-believing public. At the start of 2021, however, news organizations had started to cover legitimate concerns about whether the virus had emerged from an official facility or not. The BBC News coverage spanned both of these ends, producing an article in April 2020 that framed the lab leak theory as conspiratorial and baseless (Sardarizadeh and Robinson, 2020), and then an article just over a year later where the reporter highlighted the growing evidence backing this theory (BBC News Online, 2021a).

Chapter 6

1 A narrow focus on a specific genre is advocated by Buozis and Creech (2018, p 1438) as it provides nuance and detail that broader approaches to genre cannot.

2 Morgan Currie (2020) has done an excellent piece of work that looks to connect the concepts of performativity and performance with data. Her dissection of how Goffman's notion of performance can be used (with some caution) alongside quantification is well worth the read.

3 To understand this front-stage performance, many turn to content analysis to provide a statistical overview of how a shared set of codes emerge in news content (Mellado, 2019). This approach is valuable but its logics run counter to the case study approach of this book.

Chapter 7

1 This was one of many predictions, estimations and projections during June and July. Savid Javid, the new Health Secretary, argued that new infections could easily rise above 100,000 a day (Walker, 2021a). On the same day, *The Guardian* estimated that there would be two million UK summer COVID-19 cases (Walker, 2021b). Neil Ferguson, epidemiologist at Imperial College London, predicted that it was 'almost inevitable' that cases would reach 100,000 after the unlocking of society (Grover, 2021).

2 This is a print online publication, released every Sunday.

3 These numbers were taken from the UK government's (2021) dashboard. They refer to cases by date of specimen rather than by date reported (that is, when the actual coronavirus test was conducted).

4 Christina Pagel herself also recognized her poor projection in a series of tweets on 22 July. She explained that 'cases this week have been bit lower than many expected (inc me!)' and 'I'll join SAGE in not even trying to predict actual peak since too many uncertainties all clashing with each other' (@chrischirp, 2021).

5 It seems that projecting cases during this period was a particularly difficult art. A much more sophisticated model for future cases by SPI-M published on 7 July 2021 also overestimated the number of cases. They estimated that cases per day would be around 60,000 by 19 July – around 14,000 to 20,000 higher than the actual cases (depending on whether you use specimen or reported data) (SAGE, 2021b).

6 A caveat should be added here. The government did not introduce all the stringent measures suggested by the Imperial modelling until 23 March 2020. This delay could have caused the deaths to surpass the projection as quickly as it did.

7 We can see this in numbers that are not projections either. For example, the *Financial Times* covered an unpublished study from the University of Oxford that claimed up to 68

per cent of the UK population would have been infected with SARS-CoV-2 by 19 March 2020 (Cookson, 2020). Despite this percentage proving to be a colossal overestimate, the same story was covered by *NY Post* (Schrader, 2020), *The Standard Online* (Morrison, 2020) and *The Daily Mail Online* (May and Middleton, 2020) (among many others).

8 This data was from the government dashboard. At the time of Pagel's presentation it referred to all those vaccinated aged 18 and over (adults). This figure changed as vaccinations were rolled out.

9 These estimates are taken from Professor Ferguson's interview on Radio 4's *Today* programme on 6 July 2021. These numbers are expressed precisely but there is considerable uncertainty involved in this data. Kit Yates, in a weekly briefing by Independent SAGE (2021b), explained that determining hospitalizations and deaths from case numbers involves a series of choices by the analyst (one of which is the 'lag' between cases and hospitalizations). Therefore, they should be considered rough guides indicating the effect of the vaccine (rather than exact calculations).

10 Modelling conducted by SPI-M in early July detailed the effectiveness of the three main vaccines as part of their parameters. The effectiveness of AstraZeneca, Pfizer and Moderna against hospital admission was 90–98 per cent, while the two-dose protection against death was set at 95–98 per cent (SAGE, 2021b).

11 A document published by SAGE on 30 July outlined that it was a realistic possibility that there would be a new variant that causes 10–35 per cent case fatality and a realistic possibility that a variant emerges that evades vaccines (SAGE, 2021a).

12 See further reading on deprivation linked to (a) cases, hospitalizations and deaths (Marmot et al, 2021; Suleman et al, 2021); (b) long COVID (ONS, 2021a); (c) missed school days (Burgess et al, 2020); (d) vaccine uptake (ONS, 2021b).

13 When referring to deprivation, most researchers use *The English Indices of Deprivation 2019* (*IoD2019*). This measure is comprised of seven domains: (1) income; (2) employment; (3) health, deprivation and disability; (4) education, skills and training; (5) crime; (6) barriers to housing and services; and (7) living environment (MoHCLG, 2019). This is used to measure levels of deprivation of 32,844 small areas of the country (called Lower-layer Super Output Areas or LSOAs). After this is calculated, these areas are ranked and split into ten segments – called deprivation deciles. These range from 'most deprived' to 'least deprived' decile.

14 For more literature on health inequalities in society, read Anon's (2019) chapter in *Data in Society*.

15 This was borne out in the data about deaths and ethnicity. An ONS study in May 2020 showed that black males were 4.2 times more likely to die with COVID-19 than white males. This was largely explained by deprivation (occupation, comorbidities, education, household tenure and so on) but not entirely (White and Nafilyan, 2020). This was followed up by another ONS study in January 2022 that largely corroborated the previous report from May 2020 (Ahmed et al, 2022).

16 Doran and Cookson (2019, p 278) refer to Townsend and Davidson (1982) who state that three causes of health inequality are structural (wealth, human capital and power), organizational (inconsistent access to education, welfare and health services) and behavioural (social norms, often driven by structural factors). Here we can see how structural reorganization of the economy influences the first and third factors.

17 A similar argument concerning 'normalization' and case numbers was given by Stephen Reicher. In October 2021, as the seven-day rolling average of deaths stayed above 100 per day, he argued that the government was 'systematically normalising' the UK's current levels of infections (and resulting hospitalizations and deaths) (Devlin, 2021). Here Reicher touches on the central argument of this chapter, but does not quite go far enough. It is not just a case of normalizing high levels of infections, its about normalizing the unequal

effect of these infections and how this reflects a normalization of health inequalities in general.

18 It should be noted that Independent SAGE did consistently push for alternative policies (for example, to keep case numbers very low or to zero). In this way, they emphasized *Protect Both*. But within this specific moment, there was a dispute over whether the 'trade-off' between health and the economy made it worth opening up society on 19 July 2021.

Chapter 8

1 It may also be useful in providing an overarching theoretical framework for the relatively small collection of literature that looks at how the quantitative is communicated *within* media systems. This work focuses on specific noteworthy examples written up in journal articles and book chapters: the measurement of poverty (Lugo-Ocando and Lawson, 2018), wealth and inequality (Ginn and Duncan-Jordan, 2019; Lansley, 2019; Rosling, 2019), military attack statistics (Tasseron and Lawson, 2020), government finances (Anstead and Chadwick, 2018), healthcare metrics (Lawson, 2021b), deaths linked to conflict (Ortega-Chavez and Lawson, 2022) and transport data (Stevens, 2019).

2 Data bounds can be seen as most similar to data engines. In his second edition of *The Data Revolution*, Kitchin (2021b, p 17) argues that data actively create 'social realities' – meaning that they do not just represent or describe social reality but are constitutive. Or, put another way, data is the engine of social reality. But even in this concept, the importance of media and communication is missed. Without putting it front-and-centre, we cannot fully grasp how certain social realities are engineered by data and others are not.

3 While 'discourse' was not used in Chapter 2, this is the closest description of the data bounds mapped out. This is especially the case when we consider Stuart Hall's definition of discourse: 'a group of statements which provide a language for talking about a particular topic at a particular historical moment' (Hall, 1997, p 44).

4 It was through this lens that Nico Carpentier explained communication at a discourse studies training session at the University of Uppsala that I attended in 2017.

5 There are other groups that are not included here: members of the public, social and political activists, celebrities and sports people, and so on.

6 The Reproduction number (R) was a case study that nearly made it into the book. In fact, it made it as far as a draft chapter. Its exclusion was not because it was an uninteresting number but because of the wealth of other interesting numbers that complemented each other better. That being said, the role of R in public discourse is pertinent. It was positioned, at different points in the pandemic, as the key figure to pay attention to: if the number was below 1, the prevalence of the virus was shrinking, whereas if it was above 1, it was increasing. The way R attempted to capture this waxing and waning of coronavirus is something that needs to be explored.

7 Exactly how many is up for debate, but the delay did cost lives in the short term.

8 It should be noted here that the UK (and its four nations) are often combined with other nation-states taking the same approach. But there are UK-specific issues that may muddy the water. For example, how much of an effect has Brexit had upon the economy during the pandemic? Has it significantly affected the GDP figures at the same time as national lockdowns impacted them?

9 We can see how the UK focused on influenza in two documents from PHE (PHE, 2014b, 2014c).

10 It should be noted here that Scotland pursued an elimination strategy as it emerged from the first lockdown in the summer of 2020. This was unsuccessful, in part due to being so close to England (who adopted a mitigation approach) and a series of re-imported cases from nationals going and returning from holidays (Sridhar, 2021).

11 Another example of the importance of historical context can be found in my work with Jairo Lugo-Ocando. We argue that the singularization of 'poverty' facilitated its quantification (Lawson and Lugo-Ocando, 2018).

12 We can also look at how Saglam (2020) uses ethnographic research conducted in 2015 in rural communities in north-east Turkey to explore conspiratorial narratives. Harambam and Aupers (2014, 2016, 2021) used a 'multi-sited ethnography' conducted in the Netherlands between 2011 and 2014 to understand the 'cultic milieu' of people expressing conspiracy theories in specific relation to David Icke. The lessons from these studies can be applied to those experiencing 'information' instead.

13 I have pointed to the way data functions semi-autonomously in my work on the 'four hour wait' data in the NHS (Lawson, 2021b). Another example we can look at is GDP. Fioramonti (2013, 2014) describes this indicator as Frankenstein's monster. He emphasizes how Simon Kuznets – the creator of GDP – later became one of its fiercest critics. But, by this time, GDP has rooted itself so deeply into society that it has become synonymous with the economy itself.

14 They did have some success in pressing the government to release details about SAGE in the early part of the pandemic. This resulted in the decision to reveal the participants of SAGE and the official documents informing, and emerging from, the meetings (Inge, 2020).

15 Mennicken and Espeland (2019, p 231) explain that activists 'use numbers as a means of denunciation and criticism'. They point to specific numbers, outlining how the 1 per cent and the 99 per cent stood in for the wealthiest in society and normal Americans, respectively (Keister, 2014). Also see Didier (2018), Samuel Boris (2014) and Rodríguez-Muñiz (2021) for examples of work looking at particularly iconic numbers.

Afterword

1 This past is still set within *Trade-Off*. We can see this in the way Rishi Sunak – during his time running to be the next leader of the Conservative Party in August 2022 – explained, 'We shouldn't have empowered the scientists in the way we did. And you have to acknowledge trade-offs from the beginning' (Badshah, 2022).

References

@chrischirp (2021) Professor Christina Pagel, *Twitter*. Available at: https://twitter.com/chrischirp/status/1418293581673074692?lang=en.

@XRebellionUK (2021) Why are we doing this?, *Twitter*. Available at: https://twitter.com/XRebellionUK/status/1464173866662739980.

Ahmed, T., Drummond, R. and Bosworth, M. (2022) *Updating ethnic contrasts in deaths involving the coronavirus (COVID-19), England: 8 December 2020 to 1 December 2021*. Available at: https://www.ons.gov.uk/peoplepopulationandcommunity/birthsdeathsandmarriages/deaths/articles/updatingethniccontrastsindeathsinvolvingthecoronaviruscovid19englandandwales/8december2020to1december2021.

Ajana, B. (2017) 'Digital health and the biopolitics of the Quantified Self', *Digital Health*. doi: 10.1177/2055207616689509.

Alba, D. (2020) Virus conspiracists elevate a new champion, *New York Times*. Available at: https://www.nytimes.com/2020/05/09/technology/plandemic-judy-mikovitz-coronavirus-disinformation.html?auth=login-email&login=email.

Allington, D., Duffy, B., Wessely, S., Dhavan, N. and Rubin, J. (2020) 'Health-protective behaviour, social media usage and conspiracy belief during the COVID-19 public health emergency', *Psychological Medicine*, 51(10), pp 1763–9. Available at: https://www.cambridge.org/core/journals/psychological-medicine/article/healthprotective-behaviour-social-media-usage-and-conspiracy-belief-during-the-covid19-public-health-emergency/A0DC2C5E27936FF4D5246BD3AE8C9163.

Anderson, C. W. (2018) *Apostles of Certainty: Data journalism and the politics of doubt*. Oxford: Oxford University Press. Available at: http://eprints.whiterose.ac.uk/132706/.

Anderson, C. W. (2020a) 'Fake news is not a virus: on platforms and their effects', *Communication Theory*, 31(1), pp 42–61. doi: 10.1093/ct/qtaa008.

Anderson, C. W. (2020b) 'Practice, interpretation, and meaning in today's digital media ecosystem', *Journalism and Mass Communication Quarterly*, 97(2), pp 342–59. doi: 10.1177/1077699020916807.

Annan, J. (2020) The human cost of delaying lockdown, *Blue Skies Research*. Available at: https://bskiesresearch.wordpress.com/2020/05/12/the-human-cost-of-delaying-lockdown/.

Anon (2019) 'Health Divides', in J. Evans, S. Ruane and H. Southall (eds) *Data in Society*. Bristol: Policy Press, pp 251–64.

Anstead, N. and Chadwick, A. (2019) 'A primary definer online: the construction and propagation of a think tank's authority on social media', *Media, Culture and Society*, 40(2), pp 246–66. doi: 10.1177/0163443717707341.

Arnold, D. (1991) *Building in Egypt: Pharaonic stone masonry*. Oxford: Oxford University Press.

Arrigo, B. and Sellers, B. (2021) *The Pre-Crime Society: Crime, culture and control in the ultramodern age*. Bristol: Bristol University Press.

Assa, J. (2019) 'Gross Domestic Power: a history of GDP as numerical rhetoric', *Annals of the Fondazione Luigi Einaudi*, 53, pp 81–96.

Avilés, L. A. (2016) 'The art of public policy statisticians: the case of the Puerto Rico Pension Reform Report', *Caribbean Studies*, 44(1), pp 47–68.

Badshah, N. (2022) Sunak says it was a mistake to 'empower scientists' during Covid pandemic, *The Guardian Online*. Available at: https://www.theguardian.com/politics/2022/aug/24/sunak-says-it-was-a-mistake-to-empower-scientists-during-covid-pandemic.

Balani, S. (2020) 'Making it Count: Resisting the authority of ignorance', in G. Blakeley (ed) *Futures of Socialism*. London: Verso, pp 150–6.

Balasubramanian, S. and Cashin, P. (2019) Gross National Happiness and Macroeconomic Indicators in the Kingdom of Bhutan, *International Monetary Fund*, Working Paper No. 2019/015. Available at: https://www.imf.org/en/Publications/WP/Issues/2019/01/17/Gross-National-Happiness-and-Macroeconomic-Indicators-in-the-Kingdom-of-Bhutan-46460.

Baraniuk, C. (2021) 'Covid-19: how the UK vaccine rollout delivered success, so far', *The BMJ*, 372(421), pp 519–539.

BBC News (2020a) *Coronavirus: daily update (18 April 2020)*. Available at: https://learningonscreen.ac.uk/ondemand/index.php/prog/15F5A029X?bcast=131814758.

BBC News (2020b) *Coronavirus: daily update (19 April 2020)*. Available at: https://learningonscreen.ac.uk/ondemand/index.php/prog/15F5A0A9?bcast=131806337.

BBC News (2020c) *Coronavirus: daily update (20 April 2020)*. Available at: https://learningonscreen.ac.uk/ondemand/index.php/prog/15F95C20X?bcast=131826781.

BBC News (2020d) *Coronavirus: daily update (21 April 2020)*. Available at: https://learningonscreen.ac.uk/ondemand/index.php/prog/15F9C04E?bcast=131826977.

BBC News (2020e) *Coronavirus: daily update (22 April 2020).* Available at: https://learningonscreen.ac.uk/ondemand/index.php/prog/15F9C075?bcast=131827077.

BBC News Online (2020) Coronavirus: concern over protective kit guidance change, *BBC News Online.* Available at: https://www.bbc.co.uk/news/uk-52335561.

BBC News Online (2021a) Covid origin: why the Wuhan lab-leak theory is being taken seriously, *BBC News Online.* Available at: https://www.bbc.co.uk/news/world-asia-china-57268111.

BBC News Online (2021b) 'Staggering' number of long Covid cases, MP warns, *BBC News Online.* Available at: https://www.bbc.co.uk/news/uk-england-oxfordshire-55668650.

BBC Panorama (2020a) *BBC statement on* Panorama, *Monday 27 April: Has the government failed the NHS?* Available at: https://www.bbc.co.uk/mediacentre/statements/panorama-mon-27-apr.

BBC Panorama (2020b) *Has the government failed the NHS?* Available at: https://www.bbc.co.uk/iplayer/episode/m000hr3y/panorama-has-the-government-failed-the-nhs.

Beaver, K. (2020) Belief among Britons in coronavirus myths, *Ipsos Mori.* Available at: https://www.ipsos.com/ipsos-mori/en-uk/belief-among-britons-coronavirus-myths.

Beer, D. (2016) *Metric Power.* London: Palgrave Macmillan.

Bell, G. (2018) 'Making life: a brief history of human-robot interaction', *Consumption Markets & Culture*, 21(1), pp 22–41. doi: 10.1080/10253866.2017.1298555.

Berman, E. P. and Hirschman, D. (2018) 'The sociology of quantification: where are we now?', *Contemporary Sociology*, 47(3), pp 257–66.

Billig, M. (2019) *More Examples, Less Theory.* Cambridge: Cambridge University Press.

Billig, M. (2021) 'Rhetorical uses of precise numbers and semi-magical round numbers in political discourse about COVID-19: examples from the government of the United Kingdom', *Discourse & Society*, 32(5), pp 1–17.

Blakeley, G. (2020) *The Corona Crash.* London: Verso.

Borden, S. L. and Tew, C. (2007) 'The role of journalist and the performance of journalism: ethical lessons from 'fake' news (seriously)', *Journal of Mass Media Ethics*, 22(4), pp 300–14.

Boris, S. (2014) 'Statistics and political violence: reflections on the social conflict in 2009 in Guadeloupe', *Partecipazione e Conflitto*, 7(2), pp 237–57.

Boseley, S. and Walker, A. (2020) Hunt begins for 'close contacts' of the two UK coronavirus cases, *The Guardian Online.* Available at: https://www.theguardian.com/world/2020/jan/31/two-people-in-britain-test-positive-for-coronavirus.

Bourouiba, L. (2016) 'A sneeze', *The New England Journal of Medicine*, 375(e15). Available at: https://www.nejm.org/doi/full/10.1056/NEJMicm1501197.

Bourouiba, L. (2021) 'Fluid dynamics of respiratory infectious diseases', *Annual Review of Biomedical Engineering*, 23, pp 547–77. doi: 10.1146/annurev-bioeng-111820-025044.

Bourouiba, L., Dehandschoewercker, E. and Bush, J. W. M. (2014) 'Violent expiratory events: on coughing and sneezing', *Journal of Fluid Mechanics*, 745, pp 537–63. doi: 10.1017/jfm.2014.88.

Bratton, B. (2021) *The Revenge of the Real: Politics for a post-pandemic world*. London: Verso.

Brennen, J. S., Simon, F. M., Howard, P. N. and Nielsen, R. K. (2020) *Types, sources, and claims of COVID-19 misinformation, Reuters Institute for the Study of Journalism*. Available at: https://reutersinstitute.politics.ox.ac.uk/types-sources-and-claims-covid-19-misinformation.

Brewster, J. (2020) *Facebook bans users from sharing debunked 'Plandemic' movie, Forbes*. Available at: https://www.forbes.com/sites/jackbrewster/2020/08/18/facebook-bans-users-from-sharing-debunked-plandemic-movie/?sh=1fdc812837cc.

Brogan, C. (2021) *Coronavirus more likely to spread inside through maskless talking than coughing, Imperial College London*. Available at: https://www.imperial.ac.uk/news/213573/coronavirus-more-likely-spread-inside-through/.

Buozis, M. and Creech, B. (2018) 'Reading news as narrative: a genre approach to journalism studies', *Journalism Studies*, 19(10), pp 1430–46. doi: 10.1080/1461670X.2017.1279030.

Burgess, S., Thomson, D. Plaister, N. and Nye. P. (2020) *Pupils in the poorest areas of the country are missing the most schooling*. Available at: https://ffteducationdatalab.org.uk/2020/10/pupils-in-the-poorest-areas-of-the-country-are-missing-the-most-schooling/.

Burn-Murdoch, J. and Giles, C. (2020) UK suffers second-highest death rate from coronavirus, *Financial Times*. Available at: https://www.ft.com/content/6b4c784e-c259-4ca4-9a82-648ffde71bf0.

Cairney, P. (2021) 'The UK government's COVID-19 policy: what does "guided by the science" mean in practice?', *Frontiers in Political Science*, 3(March), pp 1–14. doi: 10.3389/fpos.2021.624068.

Calvert, J. and Arbuthnott, G. (2021) *Failures of State: The inside story of Britain's battle with coronavirus*. London: Mudlark.

Camus, A. (1970) *The Plague*. London: Penguin Books.

Carlson, M. (2018) 'Fake news as an informational moral panic: the symbolic deviancy of social media during the 2016 US presidential election', *Information Communication and Society*, 23(3), pp 374–88. doi: 10.1080/1369118X.2018.1505934.

Carr-Hill, R. (2019a) 'Measuring Social Wellbeing', in J. Evans, S. Ruane and H. Southall (eds) *Data in Society*. Bristol: Policy Press, pp 265–76.

Carr-Hill, R. (2019b) 'Using Survey Data: towards valid estimates of poverty in the South', in J. Evans, S. Ruane and H. Southall (eds) *Data in Society*. Bristol: Policy Press, pp 79–90.

CDC (2021) *Potential exposure at work, CDC*. Available at: https://www.cdc.gov/coronavirus/2019-ncov/hcp/guidance-risk-assesment-hcp.html.

Channel 4 (2020) *It is the virus killing the economy, not the restrictions themselves, Twitter*. Available at: https://twitter.com/channel4/status/1317885887871594498?lang=en.

Charumilind, S., Greenberg, E., Lamb, S. and Singhal, S. (2020) 'COVID-19: Saving thousands of lives and trillions in livelihoods', *McKinsey & Company*, (August). Available at: https://www.mckinsey.com/industries/healthcare-systems-and-services/our-insights/covid-19-saving-thousands-of-lives-and-trillions-in-livelihoods.

Cheney-Lippold, J. (2017) *We Are Data: Algorithms and the making of our digital selves*. New York: New York University Press.

Cheng, M. (2020) *UK scientists openly question government's pandemic response, AP News*. Available at: https://apnews.com/article/health-europe-virus-outbreak-london-pandemics-3493f329ee47f187740cb98ba7a28199.

Cheng, V. C., Lau, S. K., Woo, P. C. and Yuen, K. Y. (2007) 'Severe acute respiratory syndrome coronavirus as an agent of emerging and reemerging infection', *Clinical Microbiology Reviews*, 20(4), pp 660–94. doi: 10.1128/CMR.00023-07.

Chernomas, R. and Hudson, I. (2009) 'Social murder: the long-term effects of conservative economic policy', *International Journal of Health Services*, 39(1), pp 107–21. doi: 10.2190/HS.39.1.e.

Chernomas, R. and Hudson, I. (2010) 'Inequality as a cause of social murder', *International Journal of Health Services*, 40(1), pp 61–78. doi: 10.2190/HS.40.1.d.

Cherry, L. and Capel, C. (2021) U.K. passes vaccine milestone; half of country gets first dose, *Bloomberg*. Available at: https://www.bloomberg.com/news/articles/2021-04-24/u-k-passes-vaccine-milestone-half-of-country-gets-first-dose.

Chetty, R., Friedman, J. N., Hendren, N., Stepner, M. and The Opportunity Insights Team (2020) The Economic Impacts of COVID-19: Evidence from a new public database built using private sector data, *NBER*. Available at: https://www.nber.org/papers/w27431.

Chouliaraki, L. (2010) 'Discourse and Mediation', in S. Allan (ed) *Rethinking Communication: Keywords in communication research*. Cresskill, NJ: Hampton Press.

Chuquet, N. (1882) *Le Triparty en la science des nombres*. Edited by A. Marre. Rome: Imprimérie des Sciences Mathématiques.

Clarke, P. and Plimmer, G. (2020) Inside UK care homes: why the system is failing its coronavirus test, *Financial Times*. Available at: https://www.ft.com/content/86d9807e-2a47-47b2-8dff-8ab50b16e036.

Clawson, C.C. (1994) *The Mathematical Traveler*. Cambridge, MA: Perseus Publishing.

Coburn, D. (2004) 'Beyond the income inequality hypothesis: class, neo-liberalism, and health inequalities', *Social Science and Medicine*, 58(1), pp 41–56. doi: 10.1016/S0277-9536(03)00159-X.

Coburn, D. (2009) 'Inequality and Health', in L. Panitch and C. Leys (eds) *Morbid Symptoms: Health under capitalism*. Pontypool: The Merlin Press Ltd.

Coleman, S. (2018) 'The elusiveness of political truth: from the conceit of objectivity to intersubjective judgement', *European Journal of Communication*, 33, pp 157–71.

Cookson, C. (2020) Coronavirus may have infected half of UK population – Oxford study, *Financial Times*. Available at: https://www.ft.com/content/5ff6469a-6dd8-11ea-89df-41bea055720b.

Crooks, R. (2017) 'Representationalism at work: dashboards and data analytics in urban education', *Educational Media International*, 54(4), pp 289–303. doi: 10.1080/09523987.2017.1408267.

Crooks, R. and Currie, M. (2021) 'Numbers will not save us: agonistic data practices', *Information Society*, 37(4), pp 201–13. doi: 10.1080/01972243.2021.1920081.

Crossley, J. N. (2013) 'Old-fashioned versus newfangled: reading and writing numbers, 1200–1500', *Studies in Medieval and Renaissance History*, X, pp 79–109.

Currie, M. (2020) 'Data as performance: showcasing cities through open data maps', *Big Data & Society*, 7(1). doi: 10.1177/2053951720907953.

Curtis, B. (2002) 'Surveying the social: techniques, practices, power', *Histoire Social/Social History*, 35(69), pp 83–108.

CVSBWF (2013) *Blackpool: key statistics*. Available at: http://www.cvsbwf.org/wp-content/uploads/2013/09/Blackpool-Context-June-2013-Version.pdf.

Dalton, C. and Thatcher, J. (2014) What does a critical data studies look like, and why do we care? Seven points for a critical approach to 'big data', *Space and Society*. Available at: https://www.societyandspace.org/articles/what-does-a-critical-data-studies-look-like-and-why-do-we-care.

Davies, H., Pegg, D. and Lawrence, F. (2020) Revealed: value of UK pandemic stockpile fell by 40% in six years, *The Guardian Online*. Available at: https://www.theguardian.com/world/2020/apr/12/revealed-value-of-uk-pandemic-stockpile-fell-by-40-in-six-years.

Davies, N. (2009) *Flat Earth News*. London: Vintage.

Davies, N., Atkins, G., Guerin, B. and Sodhi, S. (2020) How fit were public services for coronavirus?, *CIPFA*. Available at: https://www.institutefor government.org.uk/sites/default/files/publications/how-fit-public-services-coronavirus.pdf.

Davis, N. (2020) Rival Sage group says Covid-19 policy must be clarified, *The Guardian Online*. Available at: https://www.theguardian.com/world/2020/may/04/rival-sage-group-covid-19-policy-clarified-david-king.

Dean, G., Clarke, F. and Capalbo, F. (2016) 'Pacioli's double entry: part of an intellectual and social movement', *Accounting History Review*, 26(1), pp 5–24. doi: 10.1080/21552851.2015.1129083.

Dehaene, S. (2011) *The Number Sense: How the mind creates mathematics*. Oxford: Oxford University Press.

Department for Transport (2021) *COVID-19 domestic transport data: methodology note*. Available at: https://www.gov.uk/government/statistics/transport-use-during-the-coronavirus-covid-19-pandemic/covid-19-transport-data-methodology-note.

Deringer, W. (2018) *Calculated Values: Finance, politics, and the quantitative age*. Cambridge, MA: Harvard University Press.

Descartes, R. (1996) 'Sixth Meditation', in J. Cottingham, (ed) *Meditations on First Philosophy: With selections from the objections and replies*. Cambridge: Cambridge University Press, pp 50–62.

Desrosières, A. (1998) *The Politics of Large Numbers: A history of statistical reasoning*. Cambridge, MA: Harvard University Press.

Devlin, H. (2021) Why Britons are tolerating sky-high Covid rates – and why this may not last, *The Guardian*. Available at: https://www.the guardian.com/world/2021/oct/15/why-britons-are-tolerating-sky-high-covid-rates-and-why-this-may-not-last.

Donnelly, L. (2020) UK coronavirus death toll of 20,000 'a good outcome', says chief scientific adviser, *Telegraph*. Available at: https://www.telegraph.co.uk/news/2020/03/17/governments-hopeful-scenario-20000-people-will-die-coronavirus/.

Donnelly, L., Yorke, H. and Davies, G. (2021) Test and Trace tightens 15-minute 'close contact' rule, *The Telegraph*. Available at: https://www.tele graph.co.uk/news/2021/01/11/test-trace-changes-15-minute-close-cont act-rule/.

Doran, T. and Cookson, R. (2019) 'Re-engineering health policy research to measure equity impacts', in J. Evans, S. Ruane and H. Southall (eds) *Data in Society*. Bristol: Policy Press, pp 277–89.

Dourish, P. and Gómez Cruz, E. (2018) 'Datafication and data fiction: narrating data and narrating with data', *Big Data and Society*, 5(2), pp 1–10. doi: 10.1177/2053951718784083.

Downie, A. W., Meiklejohn, M., St Vincent, L., Rao, A. R., Sundara Babu, B. V. and Kempe, C. H. (1965) 'The recovery of smallpox virus from patients and their environment in a smallpox hospital', *Bulletin of the World Health Organization*, 33(5), pp 615–22.

Duffy, B. and Allington, D. (2020) Covid conspiracies and confusions: the impact on compliance with the UK's lockdown rules and the link with social media use, *The Policy Institute* (June). Available at: https://www.kcl.ac.uk/policy-institute/assets/covid-conspiracies-and-confusions.pdf.

Duguid, J. P. (1946) 'The size and the duration of air-carriage of respiratory droplets and droplet-nuclei', *Journal of Hygiene*, 44(6), pp 471–9. doi: 10.1017/S0022172400019288.

Dunbar, R. (2010) *How Many Friends Does One Person Need?* London: Faber and Faber.

Eldred, S. M. (2021) Coronavirus FAQ: how do the rules of 6 feet and 15 minutes apply to the Delta variant?, *WUFT*. Available at: https://www.wuft.org/news/2021/08/31/coronavirus-faq-how-do-the-rules-of-6-feet-and-15-minutes-apply-to-the-delta-variant/.

Enli, G. (2015) *Mediated Authenticity: How the media constructs reality*. New York: Peter Lang.

Errington, Sian (2020) 'Against the Austerity State', in G. Blakeley (ed) *Futures of Socialism*. London: Verso, pp 177–82.

Espeland, W. and Sauder, M. (2016) *Engines of Anxiety*. New York: Russell Sage Foundation.

Espeland, W. and Stevens, M. (2008) 'A sociology of quantification', *European Journal of Sociology*, 49(3), pp 401–36.

European Environmental Agency (2016) Historical and projected shares of old-aged people (65 years and older), *European Environmental Agency*. Available at: https://www.eea.europa.eu/data-and-maps/daviz/historic-and-projected-shares-of-1#tab-chart_1.

Evans, J. (2019) 'Adult Skills Surveys and Transnational Organisations: Globalising educational policy', in J. Evans, S. Ruane and H. Southall (eds) *Data in Society*. Bristol: Policy Press, pp 65–78.

Financial Times (2020a) Covid-19: The global crisis – in data. Charts and maps show paradoxes of a pandemic that has claimed a million lives, *Financial Times*. Available at: https://ig.ft.com/coronavirus-global-data/.

Financial Times (2020b) German GDP shrinks by 2.2% in first quarter, *Financial Times*. Available at: https://www.ft.com/content/f7507260-1407-4293-bc89-9fa04b218640.

Financial Times (2020c) US GDP falls 4.8% in worst economic decline since 2008, *Financial Times*. Available at: https://www.ft.com/content/06493320-083a-4a7d-94cc-8db581f9ed03.

Finch, D. (2018) The healthy life expectancy gap, *The Health Foundation*. Available at: https://www.health.org.uk/blogs/the-healthy-life-expectancy-gap.

Fioramonti, L. (2013) *Gross Domestic Problem: The politics behind the world's most powerful number*. London: ZED Books.

Fioramonti, L. (2014) *How Numbers Rule the World: The use and abuse of statistics in global politics*. London: ZED Books.

Fisher, M. (2009) *Capitalist Realism: Is there no alternative?* Ropley: Zero Books.

Foster, P. and Neville, S. (2020) How poor planning left the UK without enough PPE, *Financial Times*. Available at: https://www.ft.com/content/9680c20f-7b71-4f65-9bec-0e9554a8e0a7.

Foucault, M. (1980) *Power/Knowledge: Selected interviews and writings, 1972–77*. New York: Random House of Canada.

Foucault, M. (2007) *Security, Territory, Population: Lectures at the Collège de France, 1977–78*. London: Palgrave Macmillan.

Fried, A. (2011) *Pathways to Polling: Crisis, cooperation and the making of public opinion professions*. London: Routledge.

Gamage, B. et al (2005) 'Protecting health care workers from SARS and other respiratory pathogens: a review of the infection control literature', *American Journal of Infection Control*, 33(2), pp 114–21. doi: 10.1016/j.ajic.2004.12.002.

Gardner, B. and Nuki, P. (2020) Exclusive: Exercise Cygnus warned the NHS could not cope with pandemic three years ago but 'terrifying' results were kept secret, *Telegraph Online*. Available at: https://www.telegraph.co.uk/news/2020/03/28/exclusive-ministers-warned-nhs-could-not-cope-pandemic-three/.

Ginn, J. and Duncan-Jordan, N. (2019) 'The Generation Game: Ending the phoney information war between young and old', in J. Evans, S. Ruane and H. Southall (eds) *Data in Society*. Bristol: Policy Press, pp 291–302.

Godoy, M. (2020) *Coronavirus FAQ: What's the deal with the '15 minute rule'?*, *NPR*. Available at: https://www.npr.org/sections/goatsandsoda/2020/10/09/922385856/coronavirus-faq-whats-the-deal-with-the-15-minute-rule.

Goffman, E. (1959) *The Presentation of Self in Everyday Life*. New York: Anchor Books.

Graham, H. (2007) *Unequal Lives: Health and socioeconomic inequalities*. Buckingham: Open University Press.

Grainger, A. (2007) 'The influence of end-users on the temporal consistency of an international statistical process: the case of tropical forest statistics', *Journal of Official Statistics*, 23(4), pp 553–92.

Grover, N. (2021) UK Covid cases could hit 200,000 a day, says scientist behind lockdown strategy, *The Guardian Online*. Available at: https://www.theguardian.com/world/2021/jul/18/uk-covid-cases-could-hit-200000-a-day-says-neil-ferguson-scientist-behind-lockdown-strategy-england.

Gurdasani, D. (2021) Ditching England's Covid restrictions is a dangerous experiment, *The Guardian Online*. Available at: https://www.theguardian.com/commentisfree/2021/jul/07/ditching-england-covid-restrictions-dangerous-mass-experiment.

Gurdasani, D., Drury, J., Greenhalgh, T., Griffin, S., Haque, Z., Katzourakis, A. et al (2021) 'Mass infection is not an option: we must do more to protect our young', *The Lancet*, 398(10297), pp 297–8. doi: 10.1016/S0140-6736(21)01589-0.

Hall, S. (1997) 'Discourse, Power and the Subject', in S. Hall (ed) *Representation: Cultural representations and signifying practices*. Milton Keynes: Open University, pp 41–5.

Hall, S., Critcher, C., Jefferson, T., Clarke, J. and Roberts, B. (1978) *Policing the Crisis*. London: The Macmillan Press.

Hallin, D. (1986) *The Uncensored War: The media and Vietnam*. New York: Oxford University Press.

Harambam, J. and Aupers, S. (2014) 'Contesting epistemic authority: conspiracy theories on the boundaries of science', *Public Understanding of Science*, 24(4), pp 466–80.

Harambam, J. and Aupers, S. (2016) '"I am not a conspiracy theorist": relational identifications in the Dutch conspiracy milieu', *Cultural Sociology*, 11(1), pp 113–29.

Harambam, J. and Aupers, S. (2021) 'From the unbelievable to the undeniable: epistemological pluralism, or how conspiracy theorists legitimate their extraordinary truth claims', *European Journal of Cultural Studies*, 24(4), pp 990–1008. doi: 10.1177/1367549419886045.

Hare, R. (1964) 'The transmission of respiratory infections', *Proceedings of the Royal Society of Medicine*, 57, pp 221–30. Available at: https://journals.sagepub.com/doi/pdf/10.1177/003591576405700329.

Hasell, J. (2020) Which countries have protected both health and the economy in the pandemic?, *Our World in Data*. Available at: https://ourworldindata.org/covid-health-economy.

Health and Social Care Committee (2021) *Coronavirus: lessons learned to date*. Available at: https://committees.parliament.uk/publications/7496/documents/78687/default/.

Hearns-Branaman, J. (2016) *Journalism and the Philosophy of Truth*. New York: Routledge.

Heffer, G. (2020) Coronavirus: experts once said 20,000 deaths was a 'good outcome' – so where is the UK headed now?, *Sky News*. Available at: https://news.sky.com/story/coronavirus-experts-once-said-20-000-deaths-was-a-good-outcome-so-where-is-the-uk-headed-now-11978609.

Hermida, A. (2015) 'Nothing but the truth: redrafting the journalistic boundary of verification', in M. Carlson and S. Lewis (eds) *Boundaries of Journalism: Professionalism, practices and participation*. New York: Routledge, pp 37–50.

Higgins, P. M. (2008) *Number Story: From counting to cryptography*. London: Copernicus Books.

Holli, M. G. (2002) *The Wizard of Washington: Emil Hurja, Franklin Roosevelt, and the birth of public opinion polling*. New York: Palgrave.

HSJ (2020) *HSJ podcast: Empty beds, death numbers, and non-Covid care*. Available at: https://www.hsj.co.uk/hsj-health-check-podcast/hsj-podcast-empty-beds-death-numbers-and-non-covid-care/7027425.article.

ICL (2020) *Report 9: Impact of non-pharmaceutical interventions (NPIs) to reduce COVID-19 mortality and healthcare demand*. London: Imperial College London. Available at: https://www.imperial.ac.uk/mrc-global-infectious-disease-analysis/covid-19/report-9-impact-of-npis-on-covid-19/.

IMF (2020) World Economic Outlook, January 2020, *IMF*. Available at: https://www.imf.org/en/Publications/WEO/Issues/2020/01/20/weo-update-january2020.

Independent SAGE (2021a) The Independent SAGE Report 44: Continuing need for support measures, *Independent SAGE*. Available at: https://www.independentsage.org/wp-content/uploads/2021/06/Supprt-document-final.pdf.

Independent SAGE (2021b) What Is Independent SAGE?, *Independent SAGE*. Available at: https://www.independentsage.org/independent-sage/.

Indie SAGE (2021) indie_SAGE 02.07.21, *YouTube*. Available at: https://www.youtube.com/watch?v=vuQnvRVkwSo.

Inge, S. (2020) Independent Sage group 'is making government more open', *Research Professional News*. Available at: https://www.researchprofessionalnews.com/rr-news-uk-politics-2020-5-independent-sage-group-is-making-government-more-open/.

IPC Global Partners (2019) *Integrated Food Security Phase Classification Technical Manual Version 3.0: Evidence and standards for better food security and nutrition decisions*. Rome: FAO.

Jaewon, K. (2020) South Korean GDP shrinks 1.4% in first quarter, most since 2008, *Nikkei Asia*. Available at: https://asia.nikkei.com/Economy/South-Korean-GDP-shrinks-1.4-in-first-quarter-most-since-2008.

Jahoda, M., Lazarsfeld, P. F. and Zeisel, H. (2002) *Marienthal: The sociography of an unemployed community*. New Brunswick, NJ: Transaction.

Jarvis, C. I. et al (2020) 'Quantifying the impact of physical distance measures on the transmission of COVID-19 in the UK', *BMC Medicine*, 18, p 124. https://doi.org/10.1186/s12916-020-01597-8.

Jenkins, R., Dowsett, A. J. and Burton, A. M. (2018) 'How many faces do people know?', *Proceedings of the Royal Society*, 285(1888).

Jennison, M. (1942) 'Atomizing of mouth and nose secretions into the air as revealed by high-speed photography', in F. R. Moulton (ed) *Aerobiology*. Washington, DC: American Association for the Advancement of Science, pp 106–28.

Jerven, M. (2013) *Poor Numbers*. Ithaca, NY: Cornell University Press.

John, I. D. (1992) 'Statistics as rhetoric in psychology', *Australian Psychologist*, 27(3), pp 144–9.

John, J. (2020) Patrick Vallance says keeping coronavirus death toll below 20,000 would be 'good outcome', *Politics Home*. Available at: https://www.politicshome.com/news/article/uks-chief-scientific-adviser-says-20000-coronavirus-deaths-would-be-good-outcome.

Johnson, K. (2022) How wrongful arrests based on AI derailed 3 men's lives, *WIRED*. Available at: https://www.wired.com/story/wrongful-arrests-ai-derailed-3-mens-lives/.

Jongsma, C. (2001) 'Mathematization and Modern Science', in R. W. Howell (ed) *Mathematics in a Postmodern Age: A Christian perspective*. Grand Rapids, MI: Eerdmans Publishing Co., pp 162–92.

Karanikolos, M. and McKee, M. (2020) How comparable is COVID-19 mortality across countries?, *COVID-19 Health System*. Available at: https://analysis.covid19healthsystem.org/index.php/2020/06/04/how-comparable-is-covid-19-mortality-across-countries/.

Katchergin, O. (2015) 'How many learning disabled are there? The rhetorical power of statistics in the Israeli discourse on learning disabilities', *Open Journal of Social Sciences*, 3(9), pp 155–66.

Keister, L. A. (2014) 'The one percent', *Annual Review of Sociology*, 40(1), pp 347–67.

Kennedy, H. and Hill, R. L. (2018) 'The feeling of numbers: emotions in everyday engagements with data and their visualisation', *Sociology*, 52(4), pp 830–48. doi: 10.1177/0038038516674675.

Kennedy, H., Hill, R. L. and Aiello, G. (2016) 'The work that visualisation conventions do', *Information, Communication & Society*, 19(6), pp 715–35.

Kilyeni, A. (2013) 'The rhetoric of numbers in print advertisements for cosmetics', *Scientific Bulletin of the Politehnica University of Timisoara. Transactions on Modern Languages/Buletinul Stiintific al Universitatii Politehnica din Timisoara. Seria Limbi Moderne*, 12(1/2), pp 17–26.

Kitchin, R. (2021a) *Data Lives*. Bristol: Bristol University Press.

Kitchin, R. (2021b) *The Data Revolution: A critical analysis of big data, open data and data infrastructures*. London: SAGE.

Kitchin, R. and Dodge, M. (2011) *Code/Space: Software and everyday life*. London: The MIT Press.

Kitchin, R. and Lauriault, T. (2014) 'Towards Critical Data Studies: Charting and unpacking data assemblages and their work'. The Programmable City Working Paper 2; pre-print version of chapter to be published in J. Eckert, A. Shears and J. Thatcher (eds) *Geoweb and Big Data*. Lincoln, NE: University of Nebraska Press.

Kitchin, R. and Lauriault, T. (2018) 'Digital Data and Data Infrastructures', in J. Ash, R. Kitchin and A. Leszczynski (eds) *Digital Geographies*. London: SAGE, pp 83–94.

Koetsenruijter, W. (2018) 'Numbers in the News: More ethos than logos?', in A. Nguyen (ed) *News, Numbers and Public Opinion in a Data-Driven World*. London: Bloomsbury Publishing, pp 177–90.

Landes, D. S. (1983) *Revolution in Time: Clocks and the making of the modern world*. Cambridge, MA: Harvard University Press.

Lansley, S. (2019) 'The "Distribution Question": Measuring and evaluating trends in inequality', in J. Evans, S. Ruane and H. Southall (eds) *Data in Society*. Bristol: Policy Press, pp 187–98.

Latour, B. (1986) 'Visualization and cognition: thinking with eyes and hands', *Knowledge and Society*, 6, pp 1–40.

Lawson, B. T. (2021a) 'Hiding behind databases, institutions and actors: how journalists use statistics in reporting humanitarian crises', *Journalism Practice*, pp 1–21.

Lawson, B. T. (2021b) 'Realizing the benefits of open government data: journalists' coverage of the NHS winter crisis, 2016–17', *The Information Society*, 38(1), pp 25–35. doi: 10.1080/01972243.2021.1998274.

Lawson, B. T. and Lovatt, M. (2020) 'Rhetoric of statistics during the NHS "winter crisis"', *European Journal of Communication*, 36(2).

Lazarsfeld, P. F. (1961) 'Notes on the history of quantification in sociology–trends, sources and problems', *Isis*, 52(2), pp 277–333.

Le, H. (2019) *Vietnam to revise GDP methodology to match international norms, VN Express*. Available at: https://e.vnexpress.net/news/business/economy/vietnam-to-revise-gdp-methodology-to-match-international-norms-3969135.html.

Louikissas, Y. (2019) *All Data Are Local: Thinking critically in a data-driven society*. Cambridge, MA: MIT Press.

Lu, G. et al (2021) 'COVID-19 in Germany and China: mitigation versus elimination strategy', *Global Health Action*, 14(1). doi: 10.1080/16549716.2021.1875601.

Lu, L., Liu, Q., Du, L. and Jiang, S. (2020) 'Middle East respiratory syndrome coronavirus (MERS-CoV): challenges in identifying its source and controlling its spread', *Microbes and Infection*, 15(8–9), pp 625–9.

Lugo-Ocando, J. (2015) *Blaming the Victim*. London: Pluto Press.

Lugo-Ocando, J. and Lawson, B. T. (2018) 'Poor Numbers, Poor News', in A. Nguyen (ed) *News, Numbers and Public Opinion in a Data-Driven World*. New York: Bloomsbury Publishing, pp 62–77.

Lupton, D. (2019) *Data Selves*. London: Polity Press.

Lynch, M. (1988) 'The externalized retina: selection and mathematization in the visual documentation of objects in the life sciences', *Human Studies*, 11, pp 201–34.

Mack, C. D. et al (2021) *Implementation and evolution of mitigation measures, testing, and contact tracing in the National Football League, August 9–November 21, 2020*. Available at: https://www.cdc.gov/mmwr/volumes/70/wr/mm7004e2.htm?s_cid=mm7004e2_w.

Mackenzie, M., Skivington, K. and Fergie, G. (2020) '"The state they're in": unpicking fantasy paradigms of health improvement interventions as tools for addressing health inequalities', *Social Science and Medicine*, 256(April), p 113047. doi: 10.1016/j.socscimed.2020.113047.

Mamdani, M. (2007) 'The politics of naming: genocide, civil war, insurgency', *London Review of Books*, 29(5), pp 5–8.

Marmot, M., Allen, J., Boyce, T., Goldblatt, P. and Morrison, J. (2021) *Building Back Fairer in Greater Manchester: Health Equity and Dignified Lives*. Report. Available at: https://www.instituteofhealthequity.org/resources-reports/build-back-fairer-in-greater-manchester-health-equity-and-dignified-lives/build-back-fairer-in-greater-manchester-main-report.pdf.

May, L. and Middleton, J. (2020) Coronavirus could already have infected HALF the British population and been spreading in the UK since JANUARY, Oxford University study claims – as official death toll jumps record 87 in a day to 422 and confirmed cases leap by 1,427, *Mail Online*. Available at: https://www.dailymail.co.uk/news/article-8148529/Coronavirus-infected-half-British-population-say-Oxford-University-experts.html.

McCarthy, M. and Carter, R. (2004) '"There's millions of them": hyperbole in everyday conversation', *Journal of Pragmatics*, 36, pp 149–84.

McCartney, G. et al (2019) 'Theorising social class and its application to the study of health inequalities', *SSM – Population Health*, 7(October 2018), p 100315. doi: 10.1016/j.ssmph.2018.10.015.

McCloskey, D. (1987) 'Rhetoric within the citadel: statistics', in J. W. Wenzel et al (eds) *Argument and Critical Practice: Proceedings of the Fifth SCA/AFA Conference on Argumentation*. Annandale, VA: Speech Communication Association, pp 485–90.

McFarland, D. A., Lewis, K. and Goldberg, A. (2016) 'Sociology in the era of big data: the ascent of forensic social science', *The American Sociologist*, 47(1), pp 12–35.

McGuinness, A. (2020) Coronavirus: focus was on NHS over care homes early in outbreak, says minister, *Sky News*. Available at: https://news.sky.com/story/coronavirus-nhs-prioritised-over-social-care-during-early-stages-of-outbreak-minister-says-11991378.

Mellado, C. (2019) 'Journalists' professional roles and role performance', in *Oxford Research Encyclopedia: Communication*. Oxford: Oxford University Press.

Mennicken, A. and Espeland, W. N. (2019) 'What's new with numbers? Sociological approaches to the study of quantification', *Annual Review of Sociology*, 45, pp 223–45. doi: 10.1146/annurev-soc-073117-041343.

Menninger, K. (1969) *Number Words and Number Symbols*. Cambridge, MA: MIT Press.

Michell, J. (2003) 'Epistemology of measurement: the relevance of its history for quantification in the social sciences', *Social Science Information*, 42(4), pp 515–34.

Michell, J. (2021) '"The art of imposing measurement upon the mind": Sir Francis Galton and the genesis of the psychometric paradigm', *Theory and Psychology*, 32(3), pp 375–400. doi: 10.1177/09593543211017671.

Mills, C. (1959) *The Sociological Imagination*. New York: Oxford University Press.

Milman, O. (2020) Covid-19 outbreak: what do health experts mean by 'flattening the curve'?, *The Guardian Online*. Available at: https://www.theguardian.com/world/2020/mar/10/covid-19-coronavirus-flattening-the-curve.

Mitchell, J. (2001) 'Clustering and psychological barriers: the importance of numbers', *The Journal of Futures Markets*, 21(5), pp 395–428.

MoHCLG (2019) *The English Indices of Deprivation 2019 (IoD2019)*. Available at: https://assets.publishing.service.gov.uk/government/uploads/system/uploads/attachment_data/file/835115/IoD2019_Statistical_Release.pdf.

Moon, N. (1999) *Opinion Polls: History, theory and practice*. Manchester: Manchester University Press.

More or Less (2021) Will the vaccine bring back normal life? GDP and fishing, *BBC Radio 4*. Available at: https://www.bbc.co.uk/sounds/play/m000rccm.

Morrison, S. (2020) Coronavirus may have infected half of UK population, experts believe, *Evening Standard*. Available at: https://www.standard.co.uk/news/health/coronavirus-half-uk-population-oxford-university-study-finds-a4396721.html

Morton, B. (2021) Covid: England lockdown rules to end on 19 July, PM confirms, *BBC News Online*. Available at: https://www.bbc.co.uk/news/uk-57809691.

MSN (2020) COVID conspiracies: 7% of Britons think there is no hard evidence that coronavirus exists, poll suggests, *MSN*. Available at: https://www.msn.com/en-gb/news/other/covid-conspiracies-7percent-of-britons-think-there-is-no-hard-evidence-that-coronavirus-exists-poll-suggests/ar-BB15FasO.

Mügge, D. and Linsi, L. (2020) 'The national accounting paradox: how statistical norms corrode international economic data', *European Journal of International Relations*, 27(2), pp 403–427. doi: 10.1177/1354066120936339.

National Bureau of Statistics of China (2020) *Preliminary Accounting Results of GDP for the First Quarter of 2020*. Available at: http://www.stats.gov.cn/english/PressRelease/202004/t20200420_1739811.html.

Navarro, V. (2007) *Consequences for Health and Quality of Life*. London: Routledge.

Navarro, V. and Shi, L. (2001) 'The political context of social inequalities and health', *International Journal of Health Sciences*, 31(1), pp 1–21.

Nguyen, A. and Lugo-Ocando, J. (2016) 'The state of data and statistics in journalism and journalism education: issues and debates', *Journalism*, 17(1), pp 3–17.

Nguyen, A., Zhao, X., Lawson, B. T. and Jackson, D. (2021) *Reporting from a statistical chaos: journalistic lessons from the first year of COVID-19 data and science in the news*. Available at: Reporting-from-a-statistical-chaos.pdf (rss.org.uk).

NHS (2020) *Risk-scoring algorithm, NHS*. Available at: https://covid19.nhs.uk/risk-scoring-algorithm.html.

Nielsen, R. K., Fletcher, R., Newman, N., Brennen, J. S. and Howard, P. N. (2020) *Navigating the 'Infodemic': How People in Six Countries Access and Rate News and Information about Coronavirus*. Oxford: Reuters Institute. Available at: https://www.politico.eu/wp-content/uploads/2020/04/Navigating-the-Coronavirus-infodemic.pdf.

NTI and Johns Hopkins (2019) *GHS Index: Global Health Security Index*. Available at: https://www.ghsindex.org/wp-content/uploads/2019/10/2019-Global-Health-Security-Index.pdf.

Oliu-Barton, M. et al (2022) 'Elimination versus mitigation of SARS-CoV-2 in the presence of effective vaccines', *The Lancet Global Health*, 10(1), pp e142–e147. doi: 10.1016/S2214-109X(21)00494-0.

Oliver, J. (1997) 'Calculations in Medieval Europe', *Mathematics in School*, 26(3), pp 12–14.

Olsen, S. J. et al (2003) 'Transmission of the Severe Acute Respiratory Syndrome on Aircraft', *New England Journal of Medicine*, 349(25), pp 2416–22. doi: 10.1056/nejmoa031349.

ONS (2019) *Gross domestic product (GDP) QMI*. Available at: https://www.ons.gov.uk/economy/grossdomesticproductgdp/methodologies/grossdomesticproductgdpqmi#methods-used-to-produce-gross-domestic-product-gdp-data.

ONS (2020a) *A guide to the UK National Accounts: March 2020*. Available at: https://www.ons.gov.uk/economy/nationalaccounts/uksectoraccounts/methodologies/aguidetotheuknationalaccountsmarch2020#gross-domestic-product-gdp.

ONS (2020b) *Comparisons of all-cause mortality between European countries and regions: January to June 2020*. Available at: https://www.ons.gov.uk/peoplepopulationandcommunity/birthsdeathsandmarriages/deaths/articles/comparisonsofallcausemortalitybetweeneuropeancountriesandregions/januarytojune2020.

ONS (2020c) *Deaths involving COVID-19 in the care sector, England and Wales: deaths occurring up to 12 June 2020 and registered up to 20 June 2020 (provisional)*. Available at: https://www.ons.gov.uk/peoplepopulationandcommunity/birthsdeathsandmarriages/deaths/articles/deathsinvolvingcovid19inthecaresectorenglandandwales/deathsoccurringupto12june2020andregisteredupto20june2020provisional#main-points.

ONS (2020d) *Deaths registered weekly in England and Wales, provisional: week ending 24 April 2020*. Available at: https://www.ons.gov.uk/peoplepopulationandcommunity/birthsdeathsandmarriages/deaths/bulletins/deathsregisteredweeklyinenglandandwalesprovisional/weekending24april2020.

ONS (2020e) *GDP monthly estimate, UK: April 2020*. Available at: https://www.ons.gov.uk/economy/grossdomesticproductgdp/bulletins/gdpmonthlyestimateuk/april2020.

ONS (2020f) *GDP monthly estimate, UK: March 2020*. Available at: https://www.ons.gov.uk/economy/grossdomesticproductgdp/bulletins/gdpmonthlyestimateuk/march2020.

ONS (2021a) *Coronavirus and the social impacts of 'long COVID' on people's lives in Great Britain: 7 April to 13 June 2021*. Available at: https://www.ons.gov.uk/peoplepopulationandcommunity/healthandsocialcare/conditionsanddiseases/articles/coronavirusandthesocialimpactsoflongcovidonpeopleslivesingreatbritain/7aprilto13june2021.

ONS (2021b) *COVID-19 vaccination rates and odds ratios by socio-demographic group*. Available at: https://www.ons.gov.uk/peoplepopulationandcommunity/healthandsocialcare/healthinequalities/datasets/covid19vaccinationratesandoddsratiosbysociodemographicgroup.

Ortega, J. and Lawson, B. T. (2022) 'How journalists do memory work with numbers: The case of the 220,000 deaths during the Colombian conflict (1958–2012)', *Journalism*, 0(0). doi: 10.1177/14648849221075434.

Osborne, R. and Laing, D. (2020) *Music by Numbers: The use and abuse of statistics in the music industries*. Bristol: Intellect.

Pavli, A., Tsiodras, S. and Maltezou, H. (2014) 'Middle East respiratory syndrome coronavirus (MERS-CoV): Prevention in travelers', *Travel Medicine and Infectious Disease*, 12(6), pp 602–8.

Peters, J. D. (2001) '"The only proper scale of representation": the politics of statistics and stories', *Political Communication*, 18(4), pp 433–49.

PHE (2014a) *MERS-CoV Close Contact Algorithm*. Available at: https://www.gov.uk/government/uploads/system/uploads/attachment_data/file/422713/Algorithm_contact_v16.pdf.

PHE (2014b) *Pandemic Influenza Response Plan 2014*. Available at: https://assets.publishing.service.gov.uk/government/uploads/system/uploads/attachment_data/file/344695/PI_Response_Plan_13_Aug.pdf.

PHE (2014c) *Pandemic Influenza Strategic Framework*. Available at: https://assets.publishing.service.gov.uk/government/uploads/system/uploads/attachment_data/file/344696/PI_Strategic_Framework_13_Aug.pdf.

PHE (2019) *MERS-CoV Close Contact Algorithm*. Available at: https://www.gov.uk/government/uploads/system/uploads/attachment_data/file/422713/Algorithm_contact_v16.pdf.

PHE (2020a) *Considerations for acute personal protective equipment (PPE) shortages*. Available at: https://www.gov.uk/government/publications/wuhan-novel-coronavirus-infection-prevention-and-control/managing-shortages-in-personal-protective-equipment-ppe.
Note to reader: this guidance was withdrawn on 27 May 2022.

PHE (2020b) *Surveillance of influenza and other seasonal respiratory viruses in the UK Winter 2020 to 2021*. Available at: https://assets.publishing.service.gov.uk/government/uploads/system/uploads/attachment_data/file/995284/Surveillance_of_influenza_and_other_seasonal_respiratory_viruses_in_the_UK_2020_to_2021-1.pdf.

PMP News (2021) This government has given up, *PMP News*. Available at: https://vip.politicsmeanspolitics.com/2021/07/04/this-government-has-given-up/.

Porter, T. (1995) *Trust in Numbers: The pursuit of objectivity in science and public life*. Princeton, NJ: Princeton University Press.

Porter, T. (2003) 'Statistics and Statistical Methods', in T. M. Porter and D. Ross (eds) *The Cambridge History of Science: Eighteenth-century science*. Cambridge: Cambridge University Press, pp 238–50.

Poutanen, S. M. and McGeer, A. J. (2004) 'Transmission and control of SARS', *Current Infectious Disease Reports*, 6(3), pp 220–7. doi: 10.1007/s11908-004-0012-7.

Prevost, J. G. (2009) *A Total Science: Statistics in Liberal and Fascist Italy*. London: McGill-Queen's University Press.

Price, S. and Harbisher, B. (2022) 'Introduction: Power, media and the Covid-19 pandemic: framing public discourse', in S. Price and Harbisher (eds) *Power, Media and the Covid-19 Pandemic*. London: Routledge.

Pringle, J. C. et al (2020) COVID-19 in a correctional facility employee following multiple brief exposures to persons with COVID-19 – Vermont, July–August 2020, *MMWR. Morbidity and Mortality Weekly Report*. doi: 10.15585/mmwr.mm6943e1.

Puzelli, S. et al. (2013) 'Investigation of an imported case of Middle East Respiratory Syndrome Coronavirus (MERS-CoV) infection in Florence, Italy, May to June 2013', *Euro Surveillance*, 18(34). doi: 10.2807/1560-7917.es2013.18.34.20564.

Qureshi, Z., Jones, N., Temple, R., Larwood, J., Greenhalgh, T. and Bourouiba, L. (2020) *What is the evidence to support the 2-metre social distancing rule to reduce COVID-19 transmission?* Available at: https://www.cebm.net/covid-19/what-is-the-evidence-to-support-the-2-metre-social-distancing-rule-to-reduce-covid-19-transmission/.

Randall, K., Ewing, E. T., Marr, L. C., Jimenez, J. L. and Bourouiba, L. (2021) 'How did we get here: what are droplets and aerosols and how far do they go? A historical perspective on the transmission of respiratory infectious diseases', *Interface Focus*, 11(6). Available at: https://royalsocietypublishing.org/doi/10.1098/rsfs.2021.0049.

Randall, S., Coast, E. and Leone, T. (2011) 'Cultural constructions of the concept of household in sample surveys', *Population Studies*, 65(2), pp 217–29.

Raphael, D. (2015) 'The political economy of health: a research agenda for addressing health inequalities in Canada', *Canadian Public Policy*, 41(Supplement 2), pp S17–S25. doi: 10.3138/cpp.2014-084.

Revely-Calder, C. (2020) The Plandemic problem: how a dubious corona-conspiracy documentary went viral, *The Telegraph Online*. Available at: https://www.telegraph.co.uk/films/0/plandemic-former-models-corona-conspiracy-documentary-took-internet/.

Rich, E., Miah, A. and Lewis, S. (2019) 'Is digital health care more equitable? The framing of health inequalities within England's digital health policy 2010–2017', *Sociology of Health and Illness*, 41(S1), pp 31–49. doi: 10.1111/1467-9566.12980.

Ritchie, H. (2020) Google mobility trends: How has the pandemic changed the movement of people around the world?, *Our World in Data*. Available at: https://ourworldindata.org/covid-mobility-trends.

Rivas, A. and Jones, T. (2020) Fact check: viral COVID-19 video spreads misinformation, makes list of false claims, *NBC San Diego*. Available at: https://www.nbcsandiego.com/news/investigations/fact-check-viral-covid-19-video-spreads-misinformation-makes-list-of-false-claims/2324126/.

Rodríguez-Muñiz, M. (2021) *Figures of the Future: Latino civil rights and the politics of demographic change*. Princeton, NJ: Princeton University Press.

Roeh, I. and Feldman, S. (1984) 'The rhetoric of numbers in front-page journalism: how numbers contribute to the melodramatic in the popular press', *Text. Interdisciplinary Journal for the Study of Discourse*, 4(4), pp 347–68.

Roozenbeek, J. et al (2020) 'Susceptibility to misinformation about COVID-19 around the world: Susceptibility to COVID misinformation', *Royal Society Open Science*, 7(10). doi: 10.1098/rsos.201199.

Rosling, H. (2019) *Factfulness: Ten reasons we're wrong about the world – and why things are better than you think*. New York: Flatiron Books.

Ruppert, E. (2012) 'The governmental topologies of database devices', *Theory, Culture & Society*, 29(5), pp 116–36. doi: 10.1177/0263276412439428.

SAGE (2021a) *S1335_Long_term_evolution_of_SARS-CoV-2*. Available at: https://assets.publishing.service.gov.uk/government/uploads/system/uploads/attachment_data/file/1007566/S1335_Long_term_evolution_of_SARS-CoV-2.pdf.

SAGE (2021b) *SPI-M-O: Summary of further modelling of easing restrictions – Roadmap Step 4 on 19th July 2021*. Available at: https://assets.publishing.service.gov.uk/government/uploads/system/uploads/attachment_data/file/1001169/S1301_SPI-M-O_Summary_Roadmap_second_Step_4.2__1_.pdf.

Saglam, E. (2020) 'What to do with conspiracy theories? Insights from contemporary Turkey', *Anthropology Today*, 36(5), pp 18–21. doi: 10.1111/1467-8322.12606.

Sample, I. (2021) 'We are a petri dish': world watches UK's race between vaccine and virus, *The Guardian Online*. Available at: https://www.theguardian.com/world/2021/jul/02/we-are-a-petri-dish-world-watches-uks-race-between-vaccine-and-virus.

Sardarizadeh, S. and Robinson, O. (2020) Coronavirus: US and China trade conspiracy theories, *BBC News Online*. Available at: https://www.bbc.co.uk/news/world-52224331.

Sarkar, A. (2019) If you got £1 every 10 seconds, you'd be a millionaire within 4 months, *Twitter*. Available at: https://twitter.com/ayocaesar/status/1193092138101354496?lang=en.

Scales, D. C. et al (2003) 'Illness in intensive care staff after brief exposure to severe acute respiratory syndrome', *Emerging Infectious Diseases*, 9(10), pp 1205–10. doi: 10.3201/eid0910.030525.

Scambler, G. (2012) 'Health inequalities', *Sociology of Health and Illness*, 34(1), pp 130–46. doi: 10.1111/j.1467-9566.2011.01387.x.

SCB (2018) *Quarterly National Accounts Inventory: sources and methods in the Swedish National Accounts*. Available at: https://www.scb.se/contentassets/767986f40d23499facc8e6f6e5234a06/_sources-and-methods-in-the-swedish-national-accounts_.pdf.

Schrader, A. (2020) Coronavirus may have already infected half of UK, study says, *New York Post*. Available at: https://nypost.com/2020/03/24/coronavirus-may-have-already-infected-half-of-uk-study-says/.

Scott-Samuel, A. and Smith, K. E. (2015) 'Fantasy paradigms of health inequalities: Utopian thinking?', *Social Theory and Health*, 13(3–4), pp 418–36. doi: 10.1057/sth.2015.12.

Seto, W. H. et al (2003) 'Effectiveness of precautions against droplets and contact in prevention of nosocomial transmission of severe acute respiratory syndrome (SARS)', *The Lancet*, 361(May). Available at: https://www.ncbi.nlm.nih.gov/pmc/articles/PMC7112437/pdf/main.pdf.

Shepherd, I. and Hearne, G. (2019) 'Data Analytics', in J. Evans, S. Ruane and H. Southall (eds) *Data in Society*. Bristol: Policy Press, pp 35–45.

Shukman, D. (2021) Covid: masks upgrade cuts infection risk, research finds, *BBC News Online*. Available at: https://www.bbc.co.uk/news/health-57636360.

Siegel, J. D., Rhinehart, E., Jackson, M., Chiarello, L., Health Care Infection Control Practices Advisory Committee (2007) '2007 guideline for isolation precautions: preventing transmission of infectious agents in health care settings', *American Journal of Infection Control*, 35(10 Suppl 2), pp S65–164. doi: 10.1016/j.ajic.2007.10.007.

Simpson, S. and Dorling, D. (1999) 'Conclusion: Statistics and "the truth"', in D. Dorling and S. Simpson (eds) *Statistics in Society*. London: Oxford University Press, pp 414–20. Available at: http://www.dannydorling.org/wp-content/files/dannydorling_publication_id3263.pdf.

Sky News (2020a) Coronavirus: UK sees the greatest total of excess deaths in care homes, *Sky News*. Available at: https://www.youtube.com/watch?v=6iediLflFFg.

Sky News (2020b) Could the UK be heading for 1,000 coronavirus deaths a day? *Sky News*. Available at: https://www.youtube.com/watch?v=q9ZnWo6jSKc.

Sky News (2020c) COVID-19: How did the UK reach 50,000 deaths? *Sky News*. Available at: https://www.youtube.com/watch?v=5mdMqYVX2c0.

Sky News (2020d) Number of UK COVID-19 deaths could reach 10,000 within a week, *Sky News*. Available at: https://www.youtube.com/watch?v=d8iQCdFavzw.

Smith, Tom W. (1990) 'The first straw? A study of the origins of election polls', *The Public Opinion Quarterly*, 54(1), pp 21–36.

Smithson, M. (2020) Data from 45 countries show containing COVID vs saving the economy is a false dichotomy, *The Conversation*. Available at: https://theconversation.com/data-from-45-countries-show-containing-covid-vs-saving-the-economy-is-a-false-dichotomy-150533?utm_source=twitter&utm_medium=bylinetwitterbutton.

Sparrow, A. (2021) UK Covid: rates of long Covid likely to increase significantly, particularly among young, warns Chris Whitty – as it happened, *The Guardian Online*. Available at: https://www.theguardian.com/politics/live/2021/jul/06/uk-covid-latest-updates-sajid-javid-new-cases-100000-per-day-summer-boris-johnson.

Squire, P. (1988) 'Why the 1936 Literary Digest Poll failed', *The Public Opinion Quarterly*, 52(1), pp 125–33.

Sridhar, D. (2020) 'COVID-19: what health experts could and could not predict', *Nature Medicine*, 26, 1812. Available at: https://www.nature.com/articles/s41591-020-01170-z.

Sridhar, D. (2021) Devi Sridhar: The UK needs a zero-Covid strategy to prevent endless lockdowns, *The New Statesman*. Available at: https://www.newstatesman.com/science-tech/2021/01/devi-sridhar-uk-needs-zero-covid-strategy-prevent-endless-lockdowns.

Statista (2022) Cumulative downloads of the NHS COVID-19 app in England and Wales as of December 2021, *Statista*. Available at: https://www.statista.com/statistics/1190062/covid-19-app-downloads-uk/.

Steerpike (2021) Five experts who predicted daily Covid cases would hit 100,000, *The Spectator*. Available at: https://www.spectator.co.uk/article/five-experts-who-predicted-covid-cases-would-hit-100-000.

Stevens, H. (2019) 'Open data, closed government: unpacking data.gov.sg', *First Monday*, 24(4).

Stuckler, D., Reeves, A., Loopstra, R., Karanikolos, M. and McKee, M. (2017) 'Austerity and health: the impact in the UK and Europe', *European Journal of Public Health*, 27(4), pp 18–21. doi: 10.1093/eurpub/ckx167.

Suleman, M., Sonthalia, S., Webb, C., Kane, M., Finch, D. and Bibby, J. (2021) *Unequal pandemic, fairer recovery. The COVID-19 impact inquiry report.* Available at: https://www.health.org.uk/publications/reports/unequal-pandemic-fairer-recovery.

Tasseron, M. and Lawson, B. T. (2020) 'Legitimising military action through statistics and discourse in the 2014 IDF assault on Gaza', *Media, War & Conflict*, 15(2), pp 238–56.

Taylor, L., Sharma, G., Martin, A. and Jameson, S. (2020) *Data Justice and COVID-19: Global perspectives*. London: Meatspace Press.

Tear, A. and Southall, H. (2019) 'Social media data', in J. Evans, S. Ruane and H. Southall (eds) *Data in Society*. Bristol: Policy Press, pp 47–60.

Temperton, J. (2020) The Plandemic conspiracy has a wild new fan club: Facebook moms, *WIRED UK*. Available at: https://www.wired.co.uk/article/plandemic-coronavirus-conspiracy-theory.

Thuy, N. (2019) *Vietnam's GDP revision follows international practices: PM, Hanoi Times*. Available at: http://hanoitimes.vn/vietnams-gdp-revision-follows-international-practices-pm-300105.html.

Till, C. (2014) 'Exercise as labour: quantified self and the transformation of exercise into labour', *Societies*, 4(3), pp 446–62.

Tolson, A. (1996) *Mediations: Text and discourse in media studies*. London: Arnold.

Townsend, P. and Davidson, N. (1982) *Inequalities in Health: The Black Report*. Harmondsworth: Penguin Books.

Triggle, N. (2021) Covid-19: A record day for cases – what does it tell us?, *BBC Online*. Available at: https://www.bbc.co.uk/news/health-59676569.

Turner, F. (2013) *The Democratic Surround: Multimedia and American liberalism from World War II to the psychedelic sixties*. Chicago: University of Chicago Press.

Turnovsky, S. J. (2013) 'The relationship between economic growth and inequality', *New Zealand Economic Papers*, 47(2), pp 113–39. doi: 10.1080/00779954.2013.776483.

UK government (2020a) *People living in deprived neighbourhoods*. Available at: https://www.ethnicity-facts-figures.service.gov.uk/uk-population-by-ethnicity/demographics/people-living-in-deprived-neighbourhoods/latest.

UK government (2020b) *Prime Minister's statement on coronavirus (COVID-19): 17 March 2020*. Available at: https://www.gov.uk/government/speeches/pm-statement-on-coronavirus-17-march-2020.

UK government (2020c) Coronavirus press conference (3 April 2020), *YouTube*. Available at: https://www.youtube.com/watch?v=XGIq9E1xfls.

UK government (2020d) *Guidance for social or community care and residential settings on COVID-19*. Available at: https://www.gov.uk/government/publications/guidance-for-social-or-community-care-and-residential-settings-on-covid-19/guidance-for-social-or-community-care-and-residential-settings-on-covid-19.

UK government (2020e) *Prime Minister's statement on coronavirus (COVID-19): 22 March 2020*. Available at: https://www.gov.uk/government/speeches/pm-statement-on-coronavirus-22-march-2020.

UK government (2021) *Coronavirus data dashboard, Gov.uk*. Available at: https://coronavirus.data.gov.uk.

UKHSA (2021) Guidance for contacts of people with confirmed coronavirus (COVID-19) infection who do not live with the person, *UK Government*. Available at: https://www.gov.uk/government/publications/guidance-for-contacts-of-people-with-possible-or-confirmed-coronavirus-covid-19-infection-who-do-not-live-with-the-person/guidance-for-contacts-of-people-with-possible-or-confirmed-coronavirus-covid-19-infection-who-do-not-live-with-the-person.

UN (2019) *Viet Nam Launches Revised GDP, United Nations*. Available at: https://vietnam.un.org/en/28249-viet-nam-launches-revised-gdp.

van Dijk, T. (2000) 'New(s) Racism: A discourse analytical approach', in S. Cottle (ed) *Ethnic Minorities and the Media: Changing cultural boundaries*. Philadelphia: Open University Press, pp 33–49.

Van Witsen, T. (2018) 'How journalists establish trust in numbers and statistics: results from an exploratory study', *Iowa State University Summer Symposium on Science Communication, 2018: Understanding the Role of Trust and Credibility in Science Communication*. Iowa State University.

Van Witsen, T. (2019) 'How daily journalists use numbers and statistics: the case of global average temperature', *Journalism Practice*, 14(9), pp 1047–65.

Varia, M. et al (2003) 'Investigation of a nosocomial outbreak of SARS in Toronto Canada', *Canadian Medical Association Journal*, 169(4), pp 285–92.

Walker, P. (2020) Vital PPE kit, including 400,000 gowns, delayed on way to UK, *The Guardian*. Available at: https://www.theguardian.com/world/2020/apr/19/vital-ppe-equipment-including-400000-gowns-en-route-to-uk-delayed-nhs-coronavirus.

Walker, P. (2021a) Covid cases could soon rise above 100,000 a day, Javid concedes, *The Guardian Online*. Available at: https://www.theguardian.com/world/2021/jul/06/covid-cases-rise-above-100000-a-day-sajid-javid-concedes-england.

Walker, P. (2021b) Data predicts 2m UK summer Covid cases with 10m isolating, *The Guardian Online*. Available at: https://www.theguardian.com/world/2021/jul/06/data-predicting-2m-uk-summer-covid-cases-prompts-health-fears.

Wells, A. (2020a) COVID conspiracies: 7% of Britons think there is no hard evidence that coronavirus exists, poll suggests, *Yahoo! News*. Available at: https://news.yahoo.com/coronavirus-conspircy-theories-135550521.html.

Wells, A. (2020b) COVID conspiracies: 7% of Britons think there is no hard evidence that coronavirus exists, poll suggests, *AoL*. Available at: https://www.aol.co.uk/news/2020/06/18/covid-conspiracies-7-of-britons-think-there-is-no-hard-evidenc/?guccounter=1&guce_referrer=aHR0cHM6Ly93d3cuZ29vZ2xlLmNvbS8&guce_referrer_sig=AQAAAAixZkQx92Muk1cu0rspn1Y0q7dKxxK9aMsBIE80iiHkX9ud3lbmVurljoDxi9sAmPPX1Nsx.

West, G. (2017) *Scale: The universal laws of growth, innovation, sustainability, and the pace of life in organisms, cities, economies, and companies*. London: Penguin Press.

Westfall, R. (1980) *Never At Rest: A biography of Isaac Newton*. Cambridge: Cambridge University Press.

White, C. and Nafilyan, V. (2020) *Coronavirus (COVID-19) related deaths by ethnic group, England and Wales: 2 March 2020 to 10 April 2020*. Available at: https://www.ons.gov.uk/peoplepopulationandcommunity/birthsdeathsandmarriages/deaths/articles/coronavirusrelateddeathsbyethnicgroupenglandandwales/2march2020to10april2020.

Whitman, B. (2022) 'The Cultural Politics of Crisis in the UK', in S. Price and B. Harbisher (eds) *Power, Media and the Covid-19 Pandemic*. London: Routledge.

WHO (2020a) *International guidelines for certification and classification (coding) of COVID-19 as cause of death*, *WHO*. Available at: https://www.who.int/classifications/icd/Guidelines_Cause_of_Death_COVID-19.pdf.

WHO (2020b) *Rational use of personal protective equipment (PPE) for coronavirus disease (COVID-19)*. Available at: https://apps.who.int/iris/bitstream/handle/10665/331498/WHO-2019-nCoV-IPCPPE_use-2020.2-eng.pdf.

WHO et al (2020c) *Managing the COVID-19 infodemic: Promoting healthy behaviours and mitigating the harm from misinformation and disinformation*, *WHO*. Available at: https://www.who.int/news/item/23-09-2020-managing-the-covid-19-infodemic-promoting-healthy-behaviours-and-mitigating-the-harm-from-misinformation-and-disinformation.

Wolf, C., Joye, D., Smith, T. W. and Fu, Y.-C. (2016) *The SAGE Handbook of Survey Methodology*. Thousand Oaks, CA: Routledge.

Wong, T. W. et al (2004) 'Cluster of SARS among medical students exposed to single patient, Hong Kong', *Emerging Infectious Diseases*, 10(2), pp 269–76. doi: 10.3201/eid1002.030452.

Woodcock, A. (2020) Coronavirus: scientists call for action after 50-fold rise in infections in schools, *Independent*. Available at: https://www.independent.co.uk/news/uk/politics/independent-sage-coronavirus-infection-schools-b1762906.html.

Woodin, C. L. (1995) *The Landmark Method for Teaching Arithmetic*. Prides Crossing, MA: Landmark School Outreach.

Woodward, K. (2009) *Statistical Panic: Cultural politics and poetics of the emotions*. Durham, NC: Duke University Press.

Worcester, R. (1991) *British Public Opinion: A guide to the history and methodology of political opinion polling*. Hoboken, NJ: Blackwell Publishing.

Wu, S. et al (2021) 'Aggressive containment, suppression, and mitigation of covid-19: lessons learnt from eight countries', *The BMJ*, 375, e067508 (table 1). doi: 10.1136/bmj-2021-067508.

XKCD Comics (2018) *Million, billion, trillion, Twitter*. Available at: https://twitter.com/xkcdComic/status/1078519122567127041.

Young Minds (2021) *Coronavirus: impact on young people with mental health needs*. Available at: https://www.youngminds.org.uk/media/esifqn3z/youngminds-coronavirus-report-jan-2021.pdf.

Yumpu (2021) Sunday people, *Yumpu*. Available at: https://www.yumpu.com/news/en/issue/101075-sunday-people-2021-07-04.

Zuboff, S. (2019) *The Age of Surveillance Capitalism*. London: Profile Books.

Index